RECODING METAPHYSICS
The New Italian Philosophy

Edited by
GIOVANNA BORRADORI

Northwestern University Press
Evanston, IL

Northwestern University Press
Evanston, IL 60201

Printed in the United States of America

Library of Congress Cataloging-in-Publication Data

Recoding metaphysics : the new Italian philosophy / edited by Giovanna
    Borradori
        p.   cm.
    Translated from the Italian.
    Bibliography: p.
    Includes index.
    ISBN 0-8101-0799-6 (alk. paper).   ISBN 0-8101-0800-3 (pbk. : alk.
paper)
    1. Philosophy, Italian—20th century.   2. Metaphysics
—History—20th century.   I. Borradori, Giovanna.
    B3601.R4   1988   88-26014
195—dc19   CIP

Contents

ACKNOWLEDGMENTS vii

GIOVANNA BORRADORI
Introduction: Recoding Metaphysics:
Strategies of the Italian Contemporary Thought 1

UMBERTO ECO
Intentio Lectoris: The State of the Art 27

GIANNI VATTIMO
Metaphysics, Violence, Secularization 45
Toward an Ontology of Decline 63

ALDO G. GARGANI
Friction of Thought 77

MARIO PERNIOLA
Venusian *Charme* 93
Decorum and Ceremony 105

PIER ALDO ROVATTI
Maintaining the Distance 117
The Black Light 123

FRANCO RELLA
The Atopy of the Modern 137
Fabula 147

MASSIMO CACCIARI
The Problem of Representation 155

EMANUELE SEVERINO
Time and Alienation 167
The Earth and the Essence of Man 177

CONTRIBUTORS 199

NOTES 201

INDEX 223

Acknowledgments

I would like to thank the Italian Foreign Affairs Bureau for their generosity in funding the translation of the essays in this volume. I would also like to thank Silvio Marchetti and Gerlando Butti of the Italian Cultural Institute for their assistance. Finally, I would like to thank Giacomo Donis, Howard Rodger MacLean, and Barbara Spackman for translating the essays, with special thanks to Barbara Spackman for her editorial work on this volume.

The essays collected in this volume originally appeared in the following Italian publications and are used here with permission.

Gianni Vattimo, "Verso un'ontologia del declino," in *Al di la del soggetto: Nietzsche, Heidegger et l'ermeneutica* (Milan: Feltrinelli, 1981).

Vattimo, "Metafisica, violenza, secolarizzazione," in G. Vattimo, ed., *Filosofia '86* (Bari: Laterza, 1986).

Aldo G. Gargani, "L'attrito del pensiero," in Vattimo, ed., *Filsofia '86*.

Mario Perniola, "Lo 'Charme' Venusian," in *Transiti* (Bologna: Cappelli, 1985).

Perniola, "Decoro e Cerimonia," in *Transiti* (Bologna: Cappelli, 1985).

Pier Aldo Rovatti, "Tenere la Distanza," in *Aut Aut* 202–203 (July–October 1984).

Rovatti, "La luce Nera," in *Aut Aut* 206–207 (March–June 1985).

Franco Rella, "Il Sogno della Ragione," in *Metamorfosi: Immagini de pensiero* (Milan: Feltrinelli, 1984).

Franco Rella, "Fabula," in *La Battaglia della Verità* (Milan: Feltrinelli, 1986).

Massimo Cacciari, "Il problema della rappresentazione," in *L'angelo necessario* (Milan: Adelphi, 1985).

Emanuele Severino, "Tempo e Alienazione," in *Gli Abitatori del Tempo* (Rome: Armando, 1978).

Severino, "La terra e l'essenza dell'uomo," in *Essenza del nichilismo* (Milan: Adelphi, 1982).

GIOVANNA BORRADORI
Recoding Metaphysics:
The New Italian Philosophy

The rise of a "national" segmentation of cultural discourse is almost always linked to a condition of weakness and marginality. This can be the case either of a community in search of its national identity, or of a nation unable to retrieve its sense of identity in the continuity with tradition because its past is "untranslatable" within the new terms imposed by integration in the postcolonial and postindustrial scene. There is no need to state that, for historical and de facto reasons, this is not the type of marginal condition in which the Italian philosophical community finds itself; and it is not even the motivation for choosing an "Italian case." It is, however, an objective fact that within its most legitimate framework, that of continental philosophy, we find Italian philosophy relegated to a subliminal position: with the exception of Benedetto Croce's neo-idealism and the revision of Marxism carried out by Antonio Gramsci, from a critical point of view the Italian debate of the century remains, if not unnoticed, then at least disarticulated, episodic, and fragmentary.[1] Hence, if on the one hand the proposal of focusing the discussion on Italian philosophy wishes to emphasize its undeservedly marginal position within the international community, on the other hand this proposal is suggested by the intrinsic nature of the Italian panorama, marked in its historic development by an extreme singularity and autonomy with respect to the Franco-German paradigms upon which non-European philosophical historiography is generally based. To presuppose an eminently "Italian" character of continental philosophy thus means embarking upon a genealogical reconstruction that, starting from the present-day situation, will illuminate the background of those ideas, conceptualizations, and recurrences whose philosophical declension today seems to be precisely and exclusively of an "Italian School."

Many factors have contributed to the perpetuation of a limited and discontinuous interest in the development of twentieth-century Italian philosophy on the part of the non-European critical community. There is no doubt that the Fascist dictatorial adventure, lasting more than twenty years, with its glorification of autarchy, did not help the diffusion and reception of Italian culture abroad. In a parallel situation, German Nazism did not have equally paralyzing results thanks to a massive intellectual emigration, Jewish above all, whose internationalism not only maintained the already existing network of contacts but also considerably strengthened interest in German culture and language between the wars and during and well after the Second World War. The obstacle created by language, laden with consequences for the very accessibility of Italian literary and philosophical culture as a whole, should, therefore, be emphasized. Unlike French, which was promoted to the role of a "language of diplomacy" from the Enlightenment until the Second World War, Italian, at least after the Renaissance, was no longer an international language, except in specific disciplines such as art history and in musical terminology.

One should add that Italian emigration did not export the national idiom but rather a series of local dialects belonging to a predominantly oral tradition and having little connection with the literary sphere. The property of little more than small family communities, dialects have always been characterized as strongly antagonistic to the prospects of integration and emancipation sought after by new generations. Lacking the linguistic instrument with which to decipher their tradition, emigrants encountered not only a "material" difficulty in attempting to recover an original tradition, but also a "cultural" difficulty in recognizing what and where that tradition might be.

The problem of the diffusion of Italian as a language with respect to the impact of philosophic culture in particular has, perhaps, suffered from yet another factor that has been dominant above all during the past two decades: the clear predominance in exportation of "visual" over "verbal" messages. Whether on the plane of mass culture or that of high culture, good examples of this tendency can be found in painting, architecture, object design, high fashion, or, more generally, the design of images. If one were to deduce that this is a historic tendency in postwar Italy, one could also offer the clamorous success of neorealistic cinema as an example. Such an almost inveterate practice of thinking of Italian culture in terms of visual codes has certainly not encouraged the translation of written texts. Indeed, even in the sector of contemporary literature, with the exception of some acclaimed masters of our postwar period, the work of diffusion has certainly not ventured into areas of experimentation, not even into those areas which are by now classical.

The question of translation, however, is not merely quantitative: the loss of historical perspectives and the consequent impossibility of reconstructing the internal horizon of referents in which the discourse of contemporary Italian philosophy takes place are the primary causes of the inability to "translate" the problematics and products of this debate within the context of continental philosophy. The first and most serious of the consequences of this is the emergence of the prejudice that Italian philosophic culture does not possess genuinely "theoretical" traits but is simply a historic-critical experience. To assert the contrary, however, is possible only if one reconstructs the physiognomy and declinations of the historic radical with which present-day thinkers still continue to engage in dialogue.[2]

The Two Sides of the Debate:
Hermeneutics and the Dissolution of Metaphysics

Before starting out on the long journey of reconstructing the historical genealogy of the Italian philosophical tradition, it is important to advance a premise about the nature and disposition of today's interlocutors.[3] This anthology is concerned with illustrating, as far as possible, that part of the Italian philosophic debate which, in the interdisciplinary panorama of the humanities, particularly in the United State, is today increasingly defined by the ecumenical notion of "theory." In consequence, we have not taken into consideration those discussions, though interesting and original, which are in some way univocally referable to a single philosophic "material," such as the philosophy of language, science, politics, law, or logic. With the field limited in this way to the sphere of strictly theoretical-aesthetic pertinence, it is interesting to point out that the stakes are substantially shared: one can, in fact, refer them back to Heidegger's radical critique of the Western metaphysical tradition—in other words, to the recognition of the fundamental inadequacy of its language. Dominated by the model of the "simple presence," the metaphysical tradition from Plato to Nietzsche has identified and confused Being with beings (entities), thereby forgetting the essential ontological "difference" that distinguishes them. But going beyond metaphysics is not a simple operation, since it is not the product of a remote theoretical "error" but the rendering explicit of a "destiny" whose achieved realization is the contemporary world, prey not only to material domination, but also and above all to the cultural domination of technology.

Within this vast horizon of referents, which the Italian debate shares with many exponents of the continental scene (ranging from Hans Georg Gadamer to Karl Otto Apel, Emmanuel Lévinas, Paul Ricoeur, Gilles Deleuze, Jean-Francois Lyotard, and Jacques Derrida), there are two positions that are both preeminent and in the greatest conflict:

1. The first converges around the proposal of interpreting the Heideggerian conceptual fabric by means of philosophical hermeneutics. A conspicuous number of thinkers fall on this side, albeit with different inflections and influences, including Gianni Vattimo, Mario Perniola, Aldo Gargani, and Pier Aldo Rovatti (from the authors included in this volume) as well as Giorgio Agamben, Gianni Carchia, Carlo Sini, Alessandro Dal Lago, Armando Rigobello, and Gianfranco Dalmasso.[4] The hermeneutic concern most of these share lies in recognizing the self-referentiality that would characterize every attempt aimed at a radical overcoming of metaphysics; that is, the modalities of its language. In other words, this means acknowledging the impossibility of placing oneself "beyond" metaphysics, for to undertake its radical overcoming would lead to accepting

just that totalizing function of thought no longer held to be theoretically legitimate. It therefore ensues that it is indispensable to maintain a dialogical-conversational relation to metaphysics, and that the current task of philosophy resides precisely in the negotiation with its language. A maieutic process par excellence, philosophy in fact corresponds to an increasingly more critical consciousness of the totalizing and self-referential presuppositions on which metaphysics is founded. From this arises the absolute centrality of the hermeneutic instance—that is, the instituting of knowledge as a "system" of interpretative reference.

2. On the opposing side is the position of those who are willing to give up on the "radical" critique of the metaphysical tradition and on the construction of a thought capable of posing itself "beyond" that nihilistic-technological destiny that has characterized the concept of Being in all of Western philosophy, from Plato to Nietzsche. This position would confront the question of Being in a "constructive" sense, breaking down the limit that Heidegger himself had accepted by giving up on the writing of the third part of *Being and Time*: "Time and Being" was to have been its ambitious title, given that precisely this final section was to have faced the problem of Time in relation to Being as such, and not the problem of Being in relation to Time and the ephemeral becoming of existence. The proposition of an *ontological question of time*, in fact, defines the area in which the second side of the Italian debate moves, inaugurated by Emanuele Severino and continued, in certain aspects, by Massimo Cacciari.[5] To pose the ontological question of time thus means soliciting the necessity for a "radical" thought in that it obliges philosophy to operate on the level of the "foundations" and not the "ways" of thinking. In clear opposition to the notion of Being as interpretation, as the movement of referral from sign to sign, from symbol to symbol, and from definition to definition, Severino arrives at the logical demonstration that the "becoming" of time is a purely illusory given. Becoming does not exist because it is founded upon an original nihilistic aporia. According to Severino it is, in fact, precisely becoming—that is, the conviction that beings arise and are annihilated *in* time—which founds the secular developments of Western metaphysical tradition, from Plato and Aristotle onward. I will later clarify the terms and justifications with which he proposes a "return to Parmenides," that is, to a static and contemplative conception of pre-Platonic Being.

Hermeneutics and Historicism

It is above all toward the hermeneutic side that international criticism has turned its attention.[6] Nevertheless, simple curiosity has failed to evolve into a more authentic and scientific interest because of the difficulty of reconstructing the network of referents internal to Italian tradition within which contemporary Italian thinkers have profound roots.[7] In order to understand their real theoretical "originality" it is not sufficient to investigate the evident affinities that Italian hermeneutics shares with the continental scene. For contemporary Italian philosophy, the hermeneutic code does not, in fact, represent a deliberate choice but rather a hereditary given, ingrained in its genetic patrimony—at least since Vico aligned himself against Descartes at the dawn of the Enlightenment and modernity. Whereas France elected itself the land of rationalism, England the cradle of empiricism, and Germany the guardian of metaphysics, Italy, with historicism, withdrew into an imaginary past, abandoning the role of cultural catalyzer that during the Renaissance had placed it at the center of the European *koinē*.

The first symptoms of a "national" coagulation of Italian philosophy took place in response to the need for a *textual* confrontation with tradition: I say textual because here the tradition does not appear as a definite complex of values, behaviors, and institutions but as a composite horizon of *historical writings*. Each of these writings, moreover, is hermeneutic by its very nature in that it represents a "historical" stage of a process of continuous interpretation of a monumental past, more imagined and desired than philologically deduced, beginning with archeological fragments. Even the construction of the cultural unity of the nation took place, in Italy, within a kaleidoscopic play of historic lenses: the Risorgimento, the mid-nineteenth-century movement to unite Italy, elected the Renaissance as its antecedent; the Renaissance in turn had recognized its element of formal and political cohesion in classical thought, in turn handed down through the words of the great Hellenistic historians and masters of rhetoric. A picture within a picture, a *regressus ad infinitum* at whose end lies a historical invention, a museological obsession.[8]

The discussion in historical writing regarding writings which had preceded it is neither playful nor necessarily ironic: its figure is neither symbolic nor surreal. This is one of the most distinctive traits of the mental garment of historicism and was to remain a legacy for Italian art and culture of this century: from the metaphysical painting of de Chirico, Carrà, and Savinio to the "Novecento" movement up until that "radical quotationism" that weak thought shares with the visual experiences of neo-Mannerism and the Transavanguardia. For "historical writing," the opening to the imaginary, the dream, the fantasy, and the irrational is never abandoned to the cathartic play of free association but is always guided by a fundamental lack of innocence. The sign always already possesses formal, conceptual, historical, and cultural connotations that guide its constant movement of reference to another sign in the dynamics of the interpretation. Thus in the hermeneutic circle the grammar of the referrals from sign to sign is constructed starting from the relative predictability of a "de-sign," and not from the relative unpredictability of the "free play" (Derrida). It should be specified that the hermeneutic mechanism of the de-sign is always implicitly a project of historical-cultural disarticulation of the sign, whereas Derridean "deconstruction" is instead oriented by the "pure" free play, or the desire for discovering—in continuation—new semantic horizons of the *grammatological* difference (*différance*).

But before we can undertake any discussion of contemporaneity, we must briefly reconstruct the roots of this hermeneutic tradition of the design which, beginning with Giambattista Vico, has forged the relationship between Italian philosophy and the codes of modernity in the form of an essential incompatibility.

Vico and the Tradition of the "De-sign"

Vico's centrality with respect to the Italian hermeneutic tradition is marked by a series of positions which, in many ways, were also to be shared by Rousseau and will determine his marginal position within the rationalistic and positivistic developments of the French culture of modernity. A generation older than Rousseau, Vico was already suspended between an initial enthusiasm for and a final rejection of the new objectivistic and rationalist methodologies which led from Descartes to the experience of the Encyclopedists. Vico, like Rousseau, was to become one of the authors most quoted by the romantics and by the German idealists. Unquestioned *Altvater*—as he was designated by Goethe—of the comparative method in the fields of jurisprudential right, ethics, and mythology, he also shared with Rousseau an interest in linguistic investigation, which has often given rise to the claim that he was "an unknown giant of linguistics" as a discipline in its own right. But unlike Rousseau, who, emblematically, directed French culture toward the idea of emancipation through action, Vico convinced Italian philosophy of the priority of memory and of form as necessary metalinguistic memory, without which not even poetry can exist; he argued for the priority of rhetoric and eloquence over grammar, for the absolute supremacy of history. For Vico, reason is a stage of historical evolution which belongs to the cycle of "courses and recourses" (*corsi e ricorsi*) of human history and which is developed after perceptive intuition and imagination. Hence myth, the invisible fruit of the notion of primitive "poetic knowledge," phenomenon among phenomena, belongs to the history of mankind. The *scienza nuova* is therefore the science of human history and for whose fruitfulness the principle of the identity of truth with fact is the guarantor (*verum ipsum factum*), given that man is certainly the author of the human world—language, myths, institutions, laws. The field of historical investigation thus circumscribes the boundaries of legitimacy of knowledge: in Kantian fashion, it renders the critical function of knowledge explicit.

History represents the defense against Cartesian ratio: Vico's historicism imprinted a profound antirationalist bias on later Italian philosophy, from Croce to the present. The very concept of "weak thought," the latest result of Italian hermeneutic reflection, is defined in clear-cut opposition to the Cartesian-rationalist tradition, with which it identifies *tout court* the totalizing root of the innovative and revolutionary "ideologies" of modernity. This "weakness" (*debolezza*) of thought is in fact taken up as the product of "a rationality which must de-strengthen itself from within, cede ground, have no fear of drawing back, . . . must not remain paralyzed by the loss of the luminous, unique and stable Cartesian reference."[9] Whereas in French philosophy Descartes inaugurated the research on the new subjectivity as research on the heart of darkness of *perception*, in Italian philosophy Vico inaugurated it under the aegis of the ineluctability of the relationship between human finitude and the universality of the time of *history*. We can trace that radical difference between the two figures who opened the doors to the philosophy of the twentieth century in France and Italy, Henri Bergson and Benedetto Croce, to this seventeenth-century divergence between the two orders of the discussion—those of perception and history.

But what was it that spurred the no-longer-young Vico to take on the anti-Cartesian polemic with so much fervor? What induced him to hypostatize the necessity of that "humanistic" *modus* which the Italian tradition after him would never abandon? In his critique of Descartes, and of rationalism in general, Vico let himself be led by the Neoplatonic tradition, which even before Augustine had been initiated by Plotinus and Porphyry. This tradition came down to Vico by way of that particular understanding of Renaissance hermetic Neoplatonism which derived from oriental wisdom the idea that the root of "truth" lies only in poetry: a theme dear to Vico that was to induce him to recognize in the aesthetic "the first operation of the human mind," and to erase the epistemological boundaries betwen philosophy and philology. Yet with the *Scienza nuova* and the exposition of the theory of the courses and recourses of history, Vico, in a historicist sense, translated the static, metahistorical, and archetypal emphasis that the hermetic tradition attributed to poetic truth. In the *Scienza nuova*, poetic truth, which remains the ground of imaginative activity (spiritual activity, par excellence), is submitted to the cyclical evolution and involution of human history, so that it is no longer poetry but history that is elected the text of texts, book of books, and writing of writings. History thus takes on the indispensable role of "formal" mediator of that original spiritual essence of which poetry is the depository.

Hence the necessity of elevating interpretation to an epistemological category, the fervor of the anti-Cartesian and antirationalist polemic, and, finally, the profoundly autochthonous matrix of Italian hermeneutics which owes little, in the genealogical sense, to the contemporary German hermeneutic ontology of Heidegger and Gadamer.

In opposition to Descartes, Vico asserts that knowledge consists not in the "consciousness" of existing but in the "science" of one's own thinking, given that the way in which thought is produced remains a mystery. The primary factor is the modality in which thought states itself: form, style, and writing become textual and not metatextual events. So it is that, with Vico's *Scienza nuova*, the Italian hermeneutic tradition defines itself to an ever greater extent as the archaeological and museological examination of the relics of ancient languages and civilizations which are "textually adrift." In this respect let us read the *etimologia fantastica* that Vico suggests for the Latin substantive "scientia": "Not fortuitously does it seem that the erudite Latin substantive *scientia* has the same etymon as the adjective *scitus*, which also means beautiful [*bello*]: inasmuch as beauty consists in a just symmetry as much of the members between themselves, as in their total combining in a beautiful body, thus should science not be considered if not as the beauty of the human mind?"[10]

Vico judges as "impious" that Promethean impetus with which Descartes tries to return to the "zero degree," to the perception of the "consciousness of existing." To this objection one could counter that to acknowledge an absolute centrality of history also represents an act of impiety. However, it is important to bear in mind that for Vico historic research was not the result of atheistic reflection but of a process of *secularization* of Christian thought: the revelation of the role and the temporal power of knowledge. And here there emerges a "spiritualist" radical of Augustinian stamp which impresses a decidedly eudaemonistic character on Vico's antirationalist position. Augustinian spiritualism directs Vico's position toward the search for the equilibrium of the soul by way of the developing of internal analysis; it convinces it of the "positive" and not destructive potential of existential excavation; and it spurs it on toward the thematic formulation of the infinite in terms of "analytics of infinitude."

Side by side with the Renaissance hermetic tradition, Augustinian spiritualism represents a second aspect of Neoplatonic reflection whose influence is not limited to Vico but extends throughout the nineteenth and twentieth centuries: from the romantic return to the ancien régime tradition which characterized the thought of Antonio Rosmini, Pasquale Galuppi, Vincenzo Gioberti, and Giuseppe Mazzini during the Italian Risorgimento; to the conjugation of existentialism with the religious problematic carried out, above all, by Luigi Pareyson; and finally to the aura of postexistentialist humanism in which the hermeneutic notion of *pietas* is located, and with which Gianni Vattimo stigmatized the rethinking of metaphysics and its world—a rethinking that can take place only under the banner of "a weak ontology, which thinks Being as trans-mission and monument . . . a patrimony constituted and handed down on the basis of prophetic illuminations."[11] Following the first historicist disinterment of Vico, by way of Risorgimento-romanticism and postwar existentialism, spiritualism has taken shape as a highly specific and distinctive trait of Italian thought.

The Choice of Historiography as the Horizon of Meaning: Benedetto Croce

The ghost of both Vico and historicism has always haunted the relationship between Italian philosophy and the quest for the "modernization" and rationalization of thought attributable to the *Aufklärung*. And this became increasingly true following the peculiar reissue of Vico's thought offered during the first twenty years of this century by Benedetto Croce. For many intellectuals of our postwar period, Vico, through Croce, was the symbol of the "closure" of the Italian academy within the nation's borders and its isolation with respect to the broader themes of the international debate. Poised between nostalgia for an ancien régime by now declining and the ferment of the new age of progress and emancipation, Vico and Croce were—and for some, still are—emblems of an endemic conservatism to be surpassed. In this violent reaction to historicism set in motion between the two world wars, which continued at least until the end of the 1950s with the contribution of Marxists, a handful of existentialists, and some phenomenologists, an important part was played by the proximity of Fascism and "neoidealistic" philosophy. Even though of the two leaders of this school, Croce and Giovanni Gentile, only the latter approved of the regime, accepting active appointments of considerable importance, Croce (who never participated in the academic structure, always playing the role of the free thinker of democratic tradition) also underwent the unjust screening of "anti-Fascist censorship." In some respects, the Fascist regime identified Croce as its "official" opponent, at times for both national and international demagogic ends. Yet notwithstanding a thorough confrontation with Marxism, Croce had always been the promoter of a bourgeois liberalism that immediately found itself in open contradiction with the emancipatory emphasis and the pervasive "politicization" of Italian left-wing culture after the Second World War. At the root of the anti-Crocean polemic, however, lie also more-strictly theoretical motivations, intrinsic to the rereading of Hegel's idealism in a "historicist" key.

The most important is an epistemological "discrimination" against the exact and social sciences, conducted on the basis of an aprioristic distinction between the "theoretical domain" (falling within the competence of art and philosophy) and the "practical domain" (the competence of ethics and economics).[12] In fact, Croce did not intend—unlike Wilhelm Dilthey and the German historicists—to confer on the human sciences (*Geisteswissenschaften*) an epistemological status equal to that given by positivism to the sciences of nature (*Naturwissenschaften*): for Croce, no such proposition of method asserted in the abstract was necessary. In his radical "antienlightenment"—which is a precise derivation from Vico's antirationalism—Croce maintained that there was no real "critical" problem of historiography in the Kantian sense, since the epistemological question, considered as a whole, is solved in the immanentistic affirmation that "life and reality are history and nothing other than history." In this perspective, which Croce himself defined as "absolute historicism," the domains of the theoretical and the practical influence each other reciprocally and are internally interdependent, with the aim of perpetuating the cyclical movement of history itself. Philosophy depends on art, which furnishes the former with the language, the means of its expression; and the practical domain depends—by its very nature—on theoretical knowledge, which illuminates it; and, finally, within the same practical domain, the economic-utilitarian moment influences the ethical moment and vice versa. History, which here is superimposed on the Hegelian notion of *Geist*, traverses its moments and its fundamental forms cyclically: each

time it proves to be enriched with the content of the previous circula-
tions, without ever repeating itself.

It is precisely this idealistic and metahistorical conception of history
that is the cause and the place of the collapse—in the Crocean system—
of that emancipatory necessity upon which the materialist vision of history is based. But the Crocean concept of history is not the major point
of friction for Marxists alone: those (few) existentialists who declared
themselves anti-Crocean also found the element of maximum incompatibility in "absolute historicism," insofar as it is improperly interpreted as
negation of the notion of subjectivity as awareness of human finitude.
And lastly, the movement of the so-called "return to reason" also aligned
itself in open antithesis to the compactness of neo-idealist philosophy
taken as a whole. In this group we might include those adepts of be-
haviorist pragmatism who, following John Dewey, inaugurated interest in
such human and social sciences as anthroplogy, linguistics, sociology, as
well as the followers of Anglo-American neopositivism, who took charge
of definitively debunking Benedetto Croce's "humanistic rhetoric." After
more than thirty-five years, we can certainly affirm now that neither the
pragmatist nor the neopositivist "opening" has consistently contributed
to the formation of "original" thought in postwar Italian philosophy.[13]

The anxiety to "internationalize" Italian culture—the professed pro-
vinciality of which was certainly to be blamed more on the Fascist autar-
chic regime than on the philosophy of Croce—led to accepting excessive
simplifications. First of these was the idea that the Crocean system was
a monolithic block endowed with a formidable compactness and transparency; in reality it was the tormented fruit of half a century of second
thoughts and rehabilitations stemming from Croce's encounter with the
most important currents of continental thought: ranging from *fin-de-siècle*
neo-Kantism to Marxism; to the psychology of Johann Herbart; and to
Dilthey, Simmel, and Bergson. Nor did Croce's approach to Hegel represent an episode of national conservatism, given that it was connected to
the contemporaneous "discovery" of the German thinker by Wilhelm Dilthey: and Hegel, between the two wars and after, was retrieved and made
sacred by the Hegel renaissance, or else his "existentialist renaissance,"
to which figures like Alexandre Kojève, Jean Hyppolite, and Jean Wahl
contributed. Max Weber's and Georg Simmel's elaborations of Croce's
thought should convince us of the pertinence of Croce's contribution to
the contemporary European debate. In his *Philosophie des Geldes*, Simmel
once again takes up Croce's approach to Marxism. Paradoxically, Croce's
detractors accused him of two opposite sins: in the notion of "absolute
historicism" some detected the signs of a regressive "metaphysical
character," whereas many "professional" philosophers repudiated him for
not being a "real" philosopher but instead a "critic" of culture, of literature—in other words, a moralist, a rhetorician. The contradictory nature
of these judgments testifies to the lack of ideological serenity with which
many intellectuals railed against Croce in the postwar period, when a

judgment of the guilts of the Fascist period was unjustly superimposed on the judgment of philosophy.

Benedetto Croce did not have only detractors, however: bypassing the more extensive field of the humanities (from literary history to art criticism and the philosophy of jurisprudential right), where one can say that his influence coincided with the "sentiment of culture" of an entire age, his impact was also crucial in the more specific field of philosophy. Together with Dilthey and Hermann Nohl, Croce was among the first in Europe of the 1910s to emphasize the centrality of Hegel's thought with respect to the formation of contemporary consciousness and subjectivity. This centrality, already glimpsed by Antonio Labriola and subsequently emphasized by Gramsci, stands at the origin of the antiscientistic and anti-Engels revision of Marxism and of the revaluation of the young Marx as opposed to the Marx of *Capital* and at the same time, in France and Italy, inaugurated that Hegel renaissance which was a fundamental ingredient of existentialism.

Thus in France, at the close of the 1960s, Jean Hyppolite acknowledged that practically from the Second World War on it was in the pages of Croce that the existentialist generation was initiated to Hegel (to the "romantic and mystical" theoretician of unhappy consciousness, to the revolutionary and not conservative Hegel). About ten years before, in Italy, Enzo Paci attributed to Croce a founding definition of life, not completely identifiable with the Hegelian notion of spirit, and therefore in correspondence with the "negative" sensibility of existentialism. As Paci wrote:

> Croce's last writings seem to be the introduction to a new Crocean philosophy, a critical rethinking of the path run, a new interpretation of the relations of Croceanism with Vico and Hegel. Vico is no longer only the philosopher of art but also the philosopher of "vitality" as barbarism, "of the terrible force of vitality," as Croce writes. And vitality leads Croce back to Hegel because "vitality . . . is restlessness that is never satisfied" and precisely for this reason ends up by becoming the "spring" of dialectic. . . . Vitality is positivity but within it there is, nevertheless, a "persistent negativity": vitality is also evil. . . . "When I am asked what Hegel has done I reply that he redeemed the world from evil because he justified it from evil."[14]

In Croce, Paci identifes the prophet of a new sentiment of life who recognizes vitality as "ambitious dimension," as "the original sin of reality." For nascent Italian existentialism, Croce's neo-Vichian historicism represented a primary point of reference with respect to the national tradition: none of the three "masters" of existentialism in Italy—Enzo Paci (1911–76), Nicola Abbagnano (1901–), and Luigi Pareyson (1918–)—has ever denied such a lineage.

Paci and Pareyson: Existentialism, Historicism, and Phenomenology

The arc of the development of the Italian "philosophical figure" shows substantial continuity from Vico to nineteenth-century romantic spiritualism; to Croce; to the existentialist adventure, which, starting in the 1940s will remain upon the scene as background noise to the "return of Husserl" of the 1960s; to the beginning of the debate on hermeneutics; and to the discussions concerning the crisis of Marxism. There were two versions of Italian existentialism which chiefly inscribed themselves on the contemporary scene.

The first was headed by Enzo Paci, who in his mature work joined the existentialist problematic to the phenomenology of the later Husserl and to some aspects of historic materialism. On a ground of referents comparable to the one on which Maurice Merleau-Ponty worked in France, and once again within the Italian tradition, Paci directed his research not toward the field of perception but toward the field of history, toward the relationship between subject and object mediated by a "projectuality" of a temporal order. From Vico and Croce, Paci inherited the propensity, within the concreteness and the contradictions of history, toward resolving the dialectical clash between reason and life, Being and being, reality and desire.[15] Paci did not develop a familiarity with psychoanalysis nor did he venture into the purely metaphysical dimension of a radical nihilism—conceptual horizons which define, instead, the *esprit* of French thought and the *Geist* of contemporary German thought. If his interest in history moved Paci toward Marxism, materialism, and the question of intellectual *engagement*, it also introduced him to some expressions of American neopositivism (above all to Alfred North Whitehead), in which he found theoretical terms for capturing the complexity of the relations between subject and object, philosophy and other fields of knowledge.[16]

It was the interest in formalizing the existential link between subjectivity and the world, and also the epistemological relationships between different disciplines, which led Paci to Husserl, particularly to the phenomenology of the later Husserl, which Merleau-Ponty was then rediscovering in the Louvain archive. Phenomenology for Paci was a transcendental science in a strictly methodological sense: the epistemological web of the great encyclopedia of sciences, the idea-limit of the interdisciplinary project.[17] This phenomenology, which owed much to the confrontation which Husserl himself faced during the last years of his life with the existentialist transgression of Heidegger, is not a "pure" science in the Kantian sense. The subject is not an originary *terminus ab quo*, a remote and archetypal U*r-ich*, but the product of an ineluctable historical and intersubjective intentionality: the relation between "subjectivity" and its project is consequently fundamental, often resolved in the coming to consciousness of an emancipatory desire. For the contemporary Italian debate, the return to Husserl begun by Paci during the 1960s enjoyed a role very similar to the one played by structuralism in France: the preeminence of an *inter-* and *trans*disciplinary formalistic hypostatization of the method as an end in itself. The dream of the method's self-justification animated the French structuralist experience and the Italian phenomenological one in parallel fashion; and the fear of the totalization and complete self-referential state of that method similarly animated the French poststructuralist experience and the hermeneutic and postphenomenological experience in Italy. What, then, divides them?

The "phenomenological structure" precludes the analysis of the unconscious and therefore any version of psychoanalysis, but not the analysis of the anamnesis of the historic and formal accretions of its appearing. The phenomenon, increasingly more so with respect to the structure, brings with it traces of a *belonging*: the "sign" emerges as a still-undivided entity of cultural "meaning." Now it is no longer by starting out from the subject—alone, pure, transparent, and still Cartesian—that the dissemination of the center and the dismemberment of the structure are carried out: the metaphor of the rhizome no longer represents the labyrinthine dispersion of the desiring subject but the labyrinthine dispersion of interpretations—in other words, of the historical backgrounds of the phenomenon, of the potential "literariness" proper to its surrounding world. To this extent language becomes the anamnestic horizon of interpretation, the line of demarcation between Being and being, historic memory and subjective memory, historicism and existentialism.

With the concept of "intentionality" phenomenology establishes that psychic activity must be in relation to an object or to other subjects, that consciousness is always consciousness *of* some thing. In other words, both consciousness and its object exist only within a reciprocal, intentional relation. "Weakness" or *debolezza* (Vattimo), "friction" or *attrito* (Gargani), and "distance" or *distanza* (Rovatti) are the rhetorical figures of this relation, which gradually comes to extinguish itself but is not yet interrupted.[18] Originary Being, absolute phenomenal purity, and the will to total suspension of every form of relationship with metaphysics are still metaphysical acts. The critique addressed by Italian hermeneutics to many versions of French poststructuralism (above all that of Deleuze, Foucault, and Jean Baudrillard) is that their attempt to go beyond metaphysics is still too radical, that the "glorification of simulacra" risks giving back to them that very *ontos on* proper to metaphysics.[19] Weakness, friction, and distance, on the other hand, are suggestions, or even only allusions to the necessity of maintaining a dialogue with a metaphysics that is worthy of attention precisely because it is worn and eroded by self-criticism. The modalities of this progressive attenuation of the intentional relationship between subject and object delimit the unstable space of Promethean action left to philosophy.

Within a horizon of referents clearly distinct from that of Paci we find the second version of Italian existentialism, that elaborated by Luigi Pareyson. Even today, it continues to exert a significant impact on the contemporary debate on hermeneutics. Whereas existentialism for Paci belonged to a markedly "atheistic" sphere of reflection (in tune with the historicism of Vico and Croce and subject to being comprised by and systematized within the methodological grid of phenomenology), for

Pareyson "existential finitude" is the answer to an ontological and religious question implying the possibility of a Christian existentialism.

Referring to a certain mystical element introduced by Schelling into romanticism, to Kierkegaard and Gabriel Marcel, Pareyson takes up the spiritualist tradition of Platonic and Augustinian origin, which, side by side with Vico's historicism, represents the second soul of the fundamental antirationalism of Italian thought.[20] The conviction that one must proceed to an interior analysis (that is, to an Augustinian dilatation of time) in order to grasp the omnipresent ontological problem affirms the absolute actuality of existentialism. Pareyson understands existentialism as a "philosophy of the person" which consciously assumes the impossibility of proposing any "objectifying" solution to philosophical problems.[21] The "speculative" character of philosophy lies exactly in this, in the indissoluble connection between truth and person, Being and person, and freedom and person. This "personalistic" perspective would, according to Pareyson, also remedy the negative conception of the individual in which both French and German existentialism have run aground. The renewal of the spiritualistic tradition, from which Pareyson at a certain point draws away, is absolutely not to be undertaken in an "intimistic" but in an ontological key. The "positive" character of the person as such depends on his ontological and "creatural" relationship with God, the recognition of whom is source and stimulus of the search for truth, always and only to be found in interpretation.

It is precisely the identification of this correspondence between existentialism and ontological dimension in the Italian debate between the 1970s and 1980s which renews the dominant interest in hermeneutics and opens an extremely close dialogue with the section of continental philosophy comprising the later Heidegger, Gadamer, Lévinas, Ricoeur, and Derrida.[22] The actuality of existentialism within the new ontological and hermeneutic sensibility, all directed toward an immanent criticism of the Western metaphysical tradition, continues to be formulated in Pareyson's terms; that is, as "the first present-day model of revivification of non-Enlightenment, 'romantic' . . . reason not content with the 'modest' and basically 'lazy' tasks of a theoretical, empirical, or technological reason, but not for this less critical of technological reason—in fact, more vigilant and aware."[23]

The presence of an existentialistic leitmotiv in the recent Italian hermeneutic community characterizes its sense of "hermeneutic philosophy" in a way differing considerably from the formulations of Heidegger, Lévinas, and Ricoeur. It is once again the "historical" instance that identifies the specificity of Italian philosophical thought: it is significant in this respect that Pareyson had already sought to resolve the ontological problem in terms of an "ontology of freedom," that is, of the manifestation of Being to the person in history, and not in terms of an epochal ontology or "ontological difference." Just as Pareyson's ontology of freedom is not projected onto an absolutely transcendent scenario but is found in the

milieu of interferences between the subject and its historical surroundings, so neither do the "weak ontology" (Vattimo) and the "friction of
thought" (Gargani) represent the vacuum of pure *différence*. Instead, they
portray the fullness of a Being understood as recollection and monument, memory, "that which is handed down." In this way weak ontology,
like the friction of thought, registers an always-waning intensity of the
conflict between the subject and its surrounding world: the progressive
weakening of the clash between historic projectuality and historic reality.
It is precisely this which places Italian hermeneutics closer to Gadamer
than to any other thinker, and which similarly distances it from Lévinas
and Derrida. It is again thanks to these considerations that one could develop the already suggested comparison between the kind of hermeneutic deferral proper to the "de-sign" (which is set in motion by the desire
to disarticulate the historical-cultural "meaning" of the "sign"), and the
kind of deconstructionist deferral proper to free play (which is set in motion instead by a centrifugal desire for associative dissemination).

Toward a Hermeneutics of Secularization

Beginning in the 1970s, interest in the ontological question related
to post-Heideggerian debate gained force among the generation that, unlike the existentialists, had not taken an active part in the events of the
Second World War or in the culture of postwar reconstruction. It was this
generation that had personally witnessed the debate on the status of
Marxism after Gramsci and Lukàcs (culminating in the spring of 1968) and
the "historical crisis" of Marxism, terminating in Italy with the internal division of the left and the tragic experience of terrorism. In fact, the recent
flowering of hermeneutics has often been judged—particularly in relation
to the personal experiences of individual authors who, like many French
poststructuralists, had dedicated part of their energies to political *engagement*—a reaction to the "crisis of ideology," in which the "crisis of ideology" is to be understood as an overturning of the categorical framework
of modernity. More specifically, in the crisis of ideology we also find a
crisis of legitimacy of rationality's "grounding reason"; the state of ob-
solescence in which the figure of the "organic" intellectual, along with his
social commitment, finds himself; and the loss of confidence in a general
"humanistic alternative" able to oppose the increasingly ever more extensive processes of technological rationalization, reification, and predominance.

Although the debate on Marxism and the crisis of ideology in a
broad sense are to be considered events of great importance in Italian—
and European—intellectual development of the past twenty years,
nevertheless I think it hasty, if not simplistic, not to identify (as much
criticism does not) the very deep roots that present-day hermeneutics
sinks into the most consolidated humus of Italian tradition. Even if to
criticize the totalizing supposition intrinsic to every "strong" ideological
position—including Marxism—is an essential objective of hermeneutic

reason, this does not imply that a critique of Marxism is its only objective or that Marxism is its privileged referent. Instead, the specificity of contemporary hermeneutic discussion lies in its being a sort of methodological "anamnesis" of the Italian philosophical tradition from Vico to postromantic spiritualism and twentieth-century existentialism. To acknowledge the multiple traces that root Italian hermeneutics in its national past is not only the expression of historical curiosity but also an indispensable operation aimed at the comprehension of the text. In fact, only a diachronic analysis is capable of deciphering the complexity of the intertextual references and of situating contextually the background of certain statements.

For example, the idea that reason can no longer be grounded beyond the rules and boundaries of language—a presupposition central to hermeneutic discourse—was already present in Vico and Croce if one thinks that these elements of the definition of language are once more understood in a truly "historic" sense. Furthermore, the "Italian" concept of hermeneutics is distinguished by the fact that language or writing are not granted a superior ontological status (something which takes place in different ways and to different degrees both in Heidegger and in Derrida); language, like the hermeneutic process that defines its interpretation, is accepted instead as the historical horizon within whose extension one carries out the progressive "desacralization" or *secularization* of metaphysical discourse. This is why Vattimo, before speaking of "weak thought," explored the notion of "ontology of decline." Linked to the concept of "sunset," crucial to the posthistoricist Italian and German traditions, Vattimo's formulation of decline starts from Heidegger's well-known thesis: that is, "the name 'Occident,' *Abendland*, not only designates our civilization's place in a geographical sense but names it ontologically as well insofar as *Abendland* is the land of the setting sun, of the sunset of Being."[24]

The distinctive characteristic of the new ontology is "sunset," understood as the progressive "weakening" of the persuasiveness of every univocal definition of Being. In other words, the new ontology is founded upon decline, to the extent that decline is not to be understood as the beginning of the end of a certain reality but as the unavoidable process of secularization that today defines our relationship to metaphysics.

The "weak" modality of our thinking therefore shares with its historicist ancestor the idea that language is "the crystallization of acts of the word, of modes of experience, located in the casket of death."[25] And it owes its existentialist progenitor for the idea that finitude is the "unit of measure" of human acting: that is, the idea that "casket"—which is death—"is also, finally, the source of those few rules which can help us move in existence in a way which is neither chaotic nor disordered, while knowing that we are not headed anywhere."[26] Even the key in which the connection between hermeneutics and nihilism is proposed reveals a historicist peculiarity, given that the concept of nihilism is also formulated within a methodological horizon. In Italian hermeneutics, nihilism, in fact, is reborn under the guise of "genealogical method," aimed at unmasking false values and ideological mystifications. Interesting in this respect is the flowering of Italian studies on Nietzsche in his "middle period," author of *Gay Science* and *Human, All-Too-Human*, a freethinker and critic of culture. Whereas much French criticism—from Derrida to Deleuze—has concentrated on the final phase of Nietzsche's production, attempting to attribute to him real ontological "theses" (for example, the idea of Being as difference), Italian criticism has focused instead upon the pseudoenlightenment and "experimental" Nietzsche, admirer of the "moralist" tradition from Montaigne to Voltaire, always taking care to maintain thought distant from any a priori and programmatic obstinacy.[27] This methodological adoption of the nihilistic viewpoint has also played an important role in the renewal of the Marxist categorical framework, especially as far as the question of technology is concerned.[28] Technology in fact has come to occupy a "destinal" horizon, since it no longer represents a variable of economic development but is established as the "destiny," both congenital and irremediable, of the Western metaphysical discourse. What the possibilities are of interfering with this destiny, and the modalities of the "conversation" to be held with it, are questions that still find themselves at the center of the Italian debate. As will become clear, this is the watershed that divides the hermeneutic camp from the more radical proposal of a definitive dissolution of metaphysics. The fate of "historicization" that falls to nihilism invests every element of the categorical picture of Italian hermeneutics: modes and itineraries—and not absolute concepts—give meaning to experience. More than ever before a consistent part of Italian thought today questions and reinterprets methodologically the antirationalist specificity of its national tradition. And it is in this aspect that its singularity within continental philosophy should be acknowledged. The debate on the crisis of reason and its foundationalist potentialities is, in reality, a revival of the dialogue with the *secularizing function of historicist reason* which a priori demystifies those questions concerning the "loss of referent" and the "glorification of simulacra" on which a large part of the French poststructuralist discussion is hinged.[29]

Indeed, the task of the secularizing function of reason is to discern and "interpret" formal quality, the *decorum*—or dignity—of appearances (Perniola). From this point of view both the original and the copy (the two theoretical poles of the notion of simulacrum) come to find themselves in a melting pot of pragmatic interferences, uses, habits, rituals, and "ceremonies" which confuse their respective values of authenticity and artificiality. This continuous movement of dissolving between the original and the copy distinguishes the typically aesthetic, "Venusian" *charme*—to cite Perniola—of interpretation. The most irreducible mark of Italian hermeneutics is to be found here: in the sensibility, the curiosity, and ability in manipulating the ethical and aesthetic deposit of the *civitas* which is to be found in every object—be it original or a copy. Going beyond metaphysics can come about only by way of a secularization of the new hieratic value of metaphysics itself, a process to be carried out in accordance with an "ethic of thinking."

The figures that animate Italian hermeneutics—Vattimo's "decline," Gargani's "friction," Rovatti's "distance," Perniola's *decorum*, and Rella's "thresholds" (*limina*)—represent the modalities of expression of this "ethic of thinking," insofar as the idea of an intentional relationship between subject and object has not yet been completely abandoned. The horizon of absolute transcendency within which phenomenology has inscribed the dualism between consciousness and the world is, by now, merely a recollection: it is substituted for, today, by a minute "care" vis-à-vis the relations and implicit pragmatic interferences in that relationship. Friction, distance, the rhetoric of the "margin," as well as the decline of any foundational ontology, all express the need for a pause, the need to apply a brake to the Cartesian arrogance of reason. In short, they describe a new space of mobility of memory understood as dislocation of subjectivity in time.

The construction of meaning is therefore subordinated to the ineluctable activity of memory, which once again takes on the purely "literary" character of the *studia humanitatis*. Memory corresponds to meditation, to a state of internal absorption, to the retreating from the "care" of the world. Recollection, while it scans the time of the "affabulation" and narration, lives contemporaneously within the space of *pietas*, which is both proof of wisdom and requisite of beauty. Ethics and aesthetics fuse to the point of constituting themselves as the sole founding dimension of thinking: rhetoric represents its instrument of argumentation. Relieved from the worry of posing as universal methodology, proper to linguistics of structuralist stamp, the art of eloquence becomes a simple mnemonic technique—that is, a series of technical precepts by which to systematize and classify the activity of memory. Rhetoric therefore assumes the function of "control" and the systematization of memory. Umberto Eco reconstructs its theoretical background as follows:

> When I say that a mnemotechnics is a semiotics, I use the term *semiotics* in the sense given it by Hjelmslev: a mnemotechnics is a connotative semiotics. To assert that the arts of memory are a semiotic phenomenon is little more than banal. Linking "y" with "x" in some fashion usually means using one as the signifier of the other. The fact that the signifier is frequently a mental image (a memory place can be either real or imaginary) does not change things. From Ockham to Peirce we have assumed that a mental icon or concept can be understood as a sign as well.[30]

This is what exonerates Italian semiotics from the self-referential labyrinth of structuralism. And Umberto Eco is to be credited with the explication of this difference. Semiotics is "necessarily" an art of memory given that semiotic process is in its own right a mechanism of "making present," of recollection.

"Every assertion, more than presupposing, *posits* the entities that it names; renders them present in the universe of discourse with semiotic force, even if only as the entity of a possible world . . . it [is] impossible to use an expression to make its own content disappear. If the arts of memory are semiotics, it is not possible to construct arts of forgetting on their model."[31] The impossibility of conceiving an *ars oblivionalis*, a rhetoric of oblivion in semiotic terms, is what gives rhetoric, *tout court*, the character of an art of memory. If it is impossible to forget (or to forget to remember), it is therefore inevitable to remember. Semiotics is thus to be considered a technique of systematization and classification that is immanent to language, which in turn is nothing other than the making present of the anamnestic artifice. Understood in this way, semiotics is also only a different declension of the hermeneutic position of Italian philosophy: of that tradition of de-sign inaugurated by Vico at the dawn of modernity. In attempting to define the relationship between his own hermeneutic notion of the text and the deconstructionist concept, Eco himself recognizes that "more than a parameter to use in order to validate the interpretation, the text is an object that the interpretation builds up in the course of the circular effort of validating itself on the basis of what it makes up as its result. I am not ashamed to admit that I am so defining the old and still valid hermeneutic circle."[32]

The text, like the sign, is not an almost abstract referentiality to which are addressed those infinite textual and intertextual interpretations of the user, as deconstruction and the aesthetics "of reception" sustain. Nor is it the self-referential product of coherences internal to the system of the text, as structuralism would have it. It is, instead, the double reciprocal projection of a "model reader," implicit in the text as *intentio operis*, and a model interpretation that the reader constructs starting out from his own interpretation of the model reader (*intentio lectoris*). It is therefore correct within this "moderate" perspective that the text be attributed with an autonomous and complete intentionality, an established *intentio operis*, as Augustine has already suggested, on the basis "of an examination of the text as a coherent entity" (Eco). Just as the interest of hermeneutics lies not in the total leveling of the dialectic between subject and world, or of the difference between language and historicity, but in the pragmatic residue inherited from these metaphysical dualisms, so also the interest of semiotics—according to Eco—is to defend the rights of interpretation, distinguishing between the use and the interpretation of a text. Although both use and interpretation are theoretical possibilities which in the empirical practice of reading always prove to be unpredictably combined, it is nevertheless necessary to preserve the sense of a polarity between them, which implies the sense of an identity of the text in its own right.

Necessity as the Unconsciousness of the Western Unconscious

As already mentioned briefly at the opening of this "genealogical" reconstruction of Italian thought, hermeneutics is not the only voice engaged in contemporary debate: in opposition to it we find the position of those who have not given up on the attempt to "go beyond" metaphysics by proposing a radical critique of its language.

Whereas the propositions of contemporary Italian hermeneutics are deeply rooted—and even "determined"—in a national past stretching from Vichian historicism to postromantic spiritualism, to Croce and existentialism, the proposal of going beyond metaphysics is connected to the tradition of Italian thought in a much more indirect way. It in fact takes up certain themes and modes of argumentation of medieval Scholasticism which—from a new point of view—supply the key to reviving some topics of Greek speculative thought (from Parmenides to Plato and Aristotle). This theoretical instrumentation, in part already explored by the Italian neo-Scholastic movement between the end of the 1940s and the 1950s, has been exhumed by Emanuele Severino and superimposed on the rethinking carried out by Heidegger covering the arc of Western thought from Plato to Nietzsche, thus giving Severino the linguistic and categorical references that allow him actively to confront the great questions of contemporary European debate.

It is interesting to recall some of the problems of neo-Scholastic reflection, given that in Italian postwar philosophy it followed an itinerary completely separate from that of the dominant existential and phenomenological interests of Paci and Pareyson, the "masters" of the present-day hermeneutic line supported by Vattimo, Gargani, Rella, Rovatti, Perniola, and—to some extent—Eco. In the atmosphere of general rejection of every form of absolute thought and of transcending the "radical historicity" of the human condition (which constituted the reply of European culture to Fascism and Nazism) neo-Scholasticism represented the interrogation of Christian thought with respect to the possibilities of metaphysical reason and ethical universality. At the close of the 1940s Etienne Gilson and Jacques Maritain in France and Gustavo Bontadini, among others, in Italy reproposed the philosophy of Thomas Aquinas as the indispensable starting point for a new foundation of ethics, of the relationship between faith and reason. In order to reassert ethics as absolute value (hence also metaphysical) Gustavo Bontadini (1903–)—who was Severino's teacher—set about establishing a process of extreme "essentialization" of metaphysical discourse.

Primarily in opposition to neopositivism, Bontadini wanted to reaffirm that fullness of Being which modern thought from Descartes to Kant had dissolved into the dualisms between certainty and truth, between appearance and reality. In particular, he asserted that the discourse of fullness of Being could begin once again by recuperating the threads of idealism on which, according to him, one ought to bestow the merit of having reopened the metaphysical possibility, in having reunited the antithesis between phenomenon and noumenon. The meaning of this fullness was reconstructed on the basis of the consideration—taken from Parmenides—that Being cannot be referred only to its determinate contents but emerges in opposition to non-Being.[33] And it is in this way that Being presents itself to mankind, in the form of becoming. Thomistic philosophy explains becoming as the principle of cause and effect: Aquinas proves that *omne quod movetur ab alio movetur*. But, adds Bontadini, Aquinas proves the means (*ab alio*) but does not prove that becoming effectively takes place (*movetur*) on the basis of that same thing: that becoming is not originary. Becoming, which takes place in sensuous experience, is consequently in contradiction with the logic of *logos*. But it is precisely this situation of conflict between experience and *logos*—according to Bontadini—that engenders the transcending of experience, and therefore legitimates metaphysical reason understood as the absolute discourse of Being.

Within the context of a radical secularization of the neo-Thomistic conceptual framework, Severino opposed Bontadini with a fundamental objection: that he had not resolved the contradiction intrinsic to the ontological status of becoming, for that contradiction does not consist in the more or less originary being of becoming but in the hypothesis of an alteration between Being and not Being implicit in the very idea of becoming. Becoming is the annulling of that which is the identification of Being with non-Being and, therefore, negates the principle of noncontradiction. The conviction that beings can be and can be annihilated in time is the very essence of metaphysical thought: by virtue of this conviction the West has determined its history as "nihilistic destiny." More specifically still, nihilism, in hypothesizing becoming as the possibility of beings to enter and leave Being, in reality defines the Being of beings by virtue of the—equivalent—hypothesis that they do not exist. In other words, nihilism consists in the conviction—unconscious but essential for the history of the West—that *being is nothing*, that the ontological status of each single being corresponds to a fundamental "alienation" from Being. At the foundations of the West. therefore, lies the concept of time as *temnein*, based on the separation of being from Being. More radically than for Heidegger, for whom the nihilism of metaphysics, from Plato onward, was determined by the confusion between Being and being, for Severino it is the hypothesis itself of the non-Being of things, or their death, which comes to constitute itself as his unconscious motive.

> For metaphysics, things "are." Their "Being" is their not-Being-a-Nothing. Insofar as they are, they are said to be "beings" (*enti*) or "Beings" (*esseri*). But being, *as such*, is that which *can* not-be: both in the sense that it could not-have-been or could not-be, and in the sense that it begins and ends (was not and is no longer). Metaphysics is the assenting to the not-Being of being.[34]

Greek thought represented the crossroads separating the nihilistic destiny of the West from the still-unexplored path that Severino defines as the "destiny of necessity," the unconsciousness of nihilism, or, put otherwise, the unconscious of the West's conscious. "According to

Necessity" (*kata to Khreōn*) is in fact the formula with which the most ancient Western text begins: the fragment by Anaximander, handed down
to history by Simplicius in his commentary on Aristotle's *Physics*. Necessity rears up on the crossroads: the thought that has led it and kept it
there is disseminated in the Greek language. Evident in the Greek word
khreōn is a series of instrumental determinations, such as the dimension
of "using" (*khraomai*) and the idea of the "hand" (*kheir*), so that in the most
archaic definition necessity designates "that of which one has need and
which the hand can take and give." For Anaximander, Necessity did not
represent a "material" instance at all, referable to a decision, a gesture, a
wish, but was instead the ineluctable appearing and concealing of a
being, whether concrete or mental, human or divine: "Necessity is the
very grip of the hand which holds and gives owing to its inescapable
strength." It is no longer the hand, the instrumental and technical instance of nihilistic dominion of the world, which follows and pursues
things, but the things themselves which obey the order given to them by
the hand, "according to Necessity."

The "timbre" of that destiny (of Necessity) which has always flowed
and will always flow outside of metaphysics is for Severino indefinable in
the methodological and programmatic terms in which the way of thinking
of the West is expressed. Necessity is not a project that one can pursue,
betray, or correct. Insofar as it is a fundamental ontological nexus which
binds Being to beings, Necessity is an "originary structure" of everything
immanent to the things of the world. To abandon nihilism in order to receive Necessity therefore implies the retrieval of an authentic meaning of
being, in which this meaning signifies "mirroring the meaning of being in
language and existing *according to* the mirroring." No instrument that is not
the ecstatic listening to the Necessity of Being in its appearing and its
concealing of itself can grasp the most authentic meaning of things:
metaphysics cannot be undone from within, as the *mise-en-abyme* of deconstruction suggests, nor is the disenchanted conversation with texts of
the metaphysical tradition inevitable, as hermeneutics and above all hermeneutic ontology maintain. The only guide is in the awareness that at
the basis of the West there lies a nihilistic mystification of the meaning
of being: the conviction that being is nothing. Furthermore, the mystification—which is also nihilistic—of the meaning of Necessity follows from
the mystification of the meaning of being. The determination of that
which is necessary is established by virtue of that which is not necessary.
Necessity is such by virtue of non-Necessity, that is, of freedom. According to Severino the destiny of the West descends from this equation between Necessity and freedom, which is nothing other than the coincidence of Necessity with the nihilistic abyss of death.

Freedom represents the essential alienation of every being from its
Being and keeps the "earth"—that is, the totality of mankind and things,
human and divine—separated from Necessity. Freedom is that which
isolates the earth from Totality and makes man think that it is the only

"safe region." Freedom, from the viewpoint of the destiny of Necessity, represents the ineradicable solitude of man and of things on the Earth.

The Necessity to which Anaximander alludes, on the other hand, is a "destiny" that does not represent the negation of freedom *tout court* but a territory different from that in which freedom and Necessity are identified. It is essential to "come out" from the metaphysical perspective in order to approach the "glimmer" of Necessity, the luminosity of the inevitable link that unites every being to its Being. Necessity does not represent a semantic field but the relation between any semantic fields whatsoever, and it is originary because it is absolutely free from its own negation. It is the principle of noncontradiction: "Necessity is such because the negation of Necessity is by necessity self-negation."[35]

Necessity thus defined inscribes itself within a totally metahistorical horizon and lies in pure contemporaneousness. Similarly, it admits of no gradation whatever by an epistemological unit of measure: its validity and operationalism are total. Thus the choice between joining or separating oneself from totality represents the watershed between the nihilistic perspective of the West and the unexplored path of Necessity. It is to the originary decision to distinguish beings from Being, the part from the whole, that the current specialistic fragmentation of knowledge should be referred. The postmetaphysical development of contemporary hermeneutics, as well as the antimetaphysical propositions advanced by neopositivism and by analytical philosophy, are the unavoidable consequences of the "decline of totality."

The immediate implication of this discourse on totality is a position of radical critique of modernity. It is precisely the thought of modernity that, in deifying the values of progress and innovation, has brought to completion that essence most proper to nihilism, which consists in the adhesion to the becoming of time understood as *temnein*, as the particularistic separating of beings from Being. To the extent that it anticipates every novelty of becoming, totality is a concept that is supremely antagonistic to the principle of modernity. Just as totality constitutes an unbreachable bastion against instrumental reason, so the decline of totality—according to Severino—leads to the absolute predominance of technology.

Greek thought is witness to the origins of this decline. "Being is and cannot Be": the definition of Being given by Parmenides glimpses the domination of totality. The completeness of Being consists in its extreme semantic simplicity, insofar as it can act neither as the subject nor as the predicate of specific beings. The question of empirical becoming put in this way, however, remains unresolved. Plato was the first and last Western philosopher to dare to go beyond Parmenides, challenging the threat of nihilism. His efforts were directed toward liberating Being from the solipsism into which Parmenides had relegated it, elevating it as the predicate of all determination. But Being (the idea), in separating into infinite determinations, fatally loses its original simplicity and is characterized as that semantic complex which is the Being-itself and not Being-other. When in the *Sophist* Plato reopens the investigation—without closing it— into the meaning of Being, he finds himself in the impasse that Severino defines as "the failed patricide." This failure by Plato to overcome Parmenides marked the definitive turn of the West toward nihilism—a turning point of which Aristotle with his doctrine of power and action became the first legislator.

The suggestion of "returning to Parmenides" is therefore not to be taken literally but rather in the sense of "beginning again from Parmenides," on an alternative path to the one taken by the West. Philosophy, as the search for the truth, is destined to remain outside of truth since "the long path along which it would claim to arrive at truth can only proceed from non-truth." The original structure of Necessity, the most authentic truth of Being, is never the starting or ending point of a search but represents instead the indispensable nexus between the transitivity of the appearing of beings on the eternal scene of Being and Being-itself.

In decreeing the illusory character of every project that aims to address, to govern, or to overthrow the necessary order of things, Severino ends up affirming the radical incompatibility between truth and individual self-determination. As dis-traction of the part from totality, subjectivity is in fact by definition always far from the truth, even when it pursues philo-sophia: "It is Necessity itself which establishes, as nontruth and as the being mortal, that being which says 'I'."[36] In this perspective, which one could perhaps call "deterministic," what space is therefore left to thought? What does thinking mean for Severino?

If for the West the action of thought has always been conceived in a dynamic sense as a gesture, an action, the purest and most crystalline of all possible gestures and actions, outside of nihilism it loses any kind of intentional characteristic and takes on the form of the "appearing of the all." Thought is the astonished wonder that arises when faced with the necessary appearing of the all, that appearing which has always already illuminated the all. Consequently, just as Necessity "is not a gesture but the place in which every gesture occurs," neither is thought a gesture but the place in which the Earth—that is, the sum total of mankind and things, human and divine—occurs. The authentic meaning of thought is a "stable circle" that delimits the eternal vault of appearing: thoughts, feelings, desires, and emotions are external stars of Being which enter and leave the vault of appearing.

According to Severino, that technological *modus* of our civilization, inaugurated at the dawn of modernity by the premises of experimental science, depends upon precisely the value of the hypothetical (or of freedom) that Western metaphysics grants to everything on Earth. Only "outside" both West and East, therefore, will it be possible to practice the authentic philo-sophia that means existing and thinking "according to necessity"—that is, according to "the appearing of the inviolable agreement between each thing and its being." Only by going back to the cross-roads in Greek thought before Plato will man be permitted to abandon the self-destructive path of nihilism and practice philo-sophia. [37]

At the end of this journey through contemporary Italian philosophy the extent to which its two major currents are inseparably tied to the vicissitudes of their national pasts appears clear. Hermeneutics developed as an intertextual discourse on the products of historicist tradition, which, starting with Vico, and by way of postromantic spiritualism, arrives at Croce and existentialism. The proposal of a radical critique of Western metaphysics, though less faithful to its own tradition, nonetheless never interrupts its dialogue with the legacy of Aristotle and the neo-Scholastic reflection, which at least from the second half of the nineteenth century has remained alive within a sector of Italian philosophy. Rooted in the evident contrast between these respective theoretical positions, the discussion between the two sides is today the most lively element of the Italian philosophical debate. Whereas hermeneutics attributes to itself a maieutic function, aimed at describing the totalizing and self-referential presuppositions on which metaphysics is founded (from which, however, a definitive detachment cannot be hypothesized), the "radical" line followed by Severino, on the other hand, does not admit the possibility of undoing the nihilism of metaphysics from within, and even less so by means of the ever more disenchanted dialogue that hermeneutics carries on with it. In this way history comes to take on a diametrically opposed role for each of the two approaches. Hermeneutics, posing itself as the composite horizon of all historical "writings," identifies in history the intertextual referent of every discursive process. On the other hand, the perspective that acknowledges the new principles of legitimacy of reason within the metahistoric sphere of Necessity fatally atrophies every dialectical function of history in which, by being self-reflective conviction that becoming appears, the nihilistic foundation of the West resides.

From the clash of these two theoretical standpoints there emerges a further fundamental disagreement concerning the attitude of philosophy toward the question of technology. If it is true for both sides that the problem of instrumental reason is to be inscribed within the broader horizon of nihilism, the ways and nature of the relationship that each side thinks philosophy must entertain with that horizon are absolutely antipodal. If for hermeneutics, in compliance with Heidegger's discussion, it is necessary to interact with the strategies of technological evolution by maintaining an ethical control over them, for those who, on the other hand, place themselves within the point of view of Necessity (foreign to the metaphysical tradition), technology remains the instrument of domination and self-destruction precisely of that nihilistic destiny in which the West finds itself trapped. This difference of positions implies a series of consequences: not only on the purely theoretical plane but also on that plane of exchanges and interrelations which both sides offer the interdisciplinary field of the humanities. The dissolution of the original dialectical density of history, which is congenital to the originary horizon of Necessity, annuls the dimension of "textuality." The destiny of Necessity, which designates the ineluctable and eternal bond which unites every being to its Being, predetermines the appearing of every reality, human and divine. The space reserved for interpretation therefore contracts to the diachronic perspective, insofar as it consists in the fundamental opposition between Western nihilism and the originary structure of Necessity. Thus philosophy closes in upon itself, conceding very little space to a confrontation of a synchronic type or to transdisciplinary contamination.

The hermeneutic line, on the other hand, in looking at history as the intertextual horizon of every discursivity, continues within the synchronic perspective to widen and intensify its incursions into the contiguous field of the humanities. Nihilism also—and above all—appears to be a polyvocal referent to "historicize," since it is not absolute concepts but the "modes" and "itineraries" of thought which constitute themselves as the meaning of experience. Thanks to the continual relaunching of the interpretive challenge, philosophy not only chooses "humanistic" contamination as its operative territory but also takes upon itself those very processes of secularization of artistic practices. Memory, understood in its broadest meaning as intertextual and metapsychological key, designates the new place of meditation. Through memory, ethics and aesthetics become mingled within a single founding dimension of thinking: philosophy becomes the "art" of arguing and interpreting by means of that "historic" instrumentation which is thought.

UMBERTO ECO
Intentio Lectoris:
The State of the Art

During the past decades we have witnessed a change of paradigm in theories of textual interpretation. In a structuralistic framework to take into account the role of the addressee looked like a disturbing intrusion since the current dogma was that a textual structure should be analyzed in itself and for the sake of itself, in an attempt to isolate its formal structures.

During the 1970s, on the other hand, literary theorists, as well as linguists and semioticians, had focused on the pragmatic aspect of reading. The dialectics between Author and Reader, Sender and Addressee, Narrator and Narratee have generated a crowd, indeed impressive, of semiotic or extrafictional narrators, subjects of the uttered utterance (*énonciation énoncée*), focalizers, voices, metanarrators, as well as an equally impressive crowd of virtual, ideal, implied or implicit, model, projected, presumed, informed readers, metareaders, archireaders, and so on.

In consequence, different critical theories, such as the aesthetics of reception, hermeneutics, the semiotic theories of interpretative cooperation, reader-response criticism, up to the scarcely homogeneous archipelago of deconstruction, have appointed as the main object of their research not so much the empirical results of given personal or collective acts of reading (studied by a sociology of reception) but the very function of construction—or deconstruction—of a text performed by its interpreter—insofar as such a function is implemented, encouraged, prescribed, or permitted by the linear textual manifestation, or by the very nature of semiosis.

It seems to me that the general assumption underlying each of these theories is that the functioning of a text (including nonverbal ones) can be explained by taking into account not only its generative process but also (or, for the most radical theories, exclusively) the role performed by the addressee and (at most) the way in which the text foresees and directs this kind of interpretative cooperation.

It must also be stressed that such an addressee-oriented approach concerns not only literary and artistic texts, but also every sort of semiosic phenomenon, comprehending everyday linguistic utterances, visual signals, and so on. In other words, addressee-oriented theories assume that the meaning of *every* message depends on the interpretative choices of its receptor: even the meaning of the most univocal message uttered in the course of the most normal communicative intercourse depends on the response of the addressee, and this response is in some way context sensitive. Naturally, such an allegedly open-ended nature of messages is more evident in those texts that have been conceived in order to magnify this semiosic possibility, that is, in so-called artistic texts. I insist on this point because during the previous decades artistic texts were taken as the only phenomenon able to display, provocatively, the still-unacknowledged open-ended nature of texts. In the past decades, however, such a nature has been theoretically rooted into the very nature of any kind of text. In other words, before the change of the paradigm, artistic texts were seen as the only cases in which a semiosic system, be it verbal or not, magnified the role of the addressee—the basic and normal function of such a system being instead that of allowing an ideal condition of univocality independent of the idiosyncrasies of the receptor. But recently semiotic theories have insisted that—even though in everyday life we are obliged to exchange many univocal messages, working (with difficulty) to reduce ambiguity—the dialectic between sender, addressee, and context is at the very core of semiosis.

This paper will focus particularly on the change of paradigm in literary theories. The reasons will be clear in the course of the next paragraphs: facing the new paradigm I shall take, courageously, a "moderate" standpoint, arguing against some degenerations of so-called reader-response criticism. I shall claim that a theory of interpretation—even when it assumes that texts are open to multiple readings—must also assume that it is possible to reach an agreement, if not about the meanings that a text encourages, at least about those that a text discourages. Since literary texts are today viewed as the most typical phenomenon of unlimited semiosis, it will be worthwhile to debate the problem of textuality there where the very notion of text seems to dissolve into a whirl of individual readings.

1. Archaeology

Undoubtedly the universe of literary studies has been haunted, in the past years, by the ghost of the reader. To prove this assumption it will be interesting to ascertain how and to what extent such a ghost has been conjured up by different theorists, coming from different theoretical traditions.

The first who explicitly spoke of an "implied author" ("carrying the reader with him") was certainly Wayne Booth (1961). After him we can isolate two independent lines of research that, until a certain moment, ignored each other, namely, the semiotico-structural one and the hermeneutic one.

The first line stems from *Communications* 8, where Roland Barthes spoke of a material author that cannot be identified with the narrator. Tzvetan Todorov evoked the coupled "image of the narrator—image of the author" and recovered Anglo-Saxon theories of point of view (from Henry James, Percy Lubbock, E. M. Forster, up to Pouillon [1946]), and Gerard Genette started to elaborate the categories (definitely dealt with in 1972) of voice and focalization. This line also includes some observations of Julia Kristeva (1970) on "textual productivity," certain lucid pages of Jurij Lotman (1977), the still-empirical concept of *architecteur* by Michael Riffaterre (1971), the discussions on the conservative standpoint of E. D. Hirsch (1967), and the debate brought to the most elaborated notions of implied reader in Maria Corti (1976) and Seymour Chatman (1978). It is interesting to remark that the last two authors drew their definition directly from Booth, ignoring the similar definition proposed by Wolfgang Iser in 1972. The same happened to me, and I elaborated my notion of Model Reader along the mainstream of the semiotic-structuralistic line, matching these results with some suggestions borrowed from various discussions on the modal logic of narrativity (mainly Tehn van Djik, Janos Petöfi, and Schmidt) as well as from some hints furnished by Weinrich—not to speak of the idea of an "ideal reader" designated by James Joyce for *Finnegans Wake*.

It is also interesting to remark that Corti (1976) traces the discussion on the nonempirical author back to Michel Foucault (1977), who, in a poststructuralist atmosphere, posits the problem of an author as a "way of being within the discourse," a field of conceptual coherence, a stylistic unity, which as such could not but elicit the corresponding idea of a reader as a way of recognizing such a being-within-the-discourse.

The second lineage is represented by Iser (1974), who starts from Booth's proposal but elaborates his suggestion on the basis of a different tradition (Roman Ingarden, Hans Georg Gadamer, and naturally Hans Robert Jauss—who in his turn was developing some of the suggestions of the Russian formalists and the Prague school). Iser was also largely influenced (as it is demonstrated by the bibliographical references of *Der implizierte Leser*) by the Anglo-Saxon theorists of narrativity (well known by Todorov and Genette) and by Joycean criticism. One finds in Iser's first book few references to the structuralist lineage (the only important source is Jan Mukarovsky). It is only in *The Act of Reading* (1978) that he brilliantly (and better informed than his structuralistic colleagues) tries to reconnect the two lineages, with references to Roman Jakobson, Lotman, Hirsch, and Riffaterre, as well as to some of my remarks of the early 1960s.

Such an insistence on the moment of reading, coming from different directions, seems to reveal a felicitous plot of the zeitgeist. And speaking of the zeitgeist, it is curious to notice that at the beginning of the 1980s Charles Fillmore, coming from the autonomous and different tradition of generative semantics (critically reviewed), wrote an essay entitled

"Ideal Readers and Real Readers" without any conscious reference to the above-mentioned debates.

Certainly all these author/reader couples do not have the same theoretical status (for a brilliant map of their mutual differences and identities see Pugliatti [1985]). However, the most important problem, it seems to me, is to ascertain whether such a reader-oriented atmosphere really represented a new trend in aesthetic and semiotic studies or not.

Sifting through the advocates of this trend, I am happy to find myself and my fellow travelers proceeding in a very respectable historical mainstream.

The entire history of aesthetics can be traced back to a history of theories of interpretation and of the effect that a work of art provokes in its addressee. I consider response-oriented Aristotle's *Poetics*, the pseudo-Longinian aesthetics of the Sublime, the medieval theories of beauty as the final result of a "visio," the new reading of Aristotle performed by the Renaissance theorists of drama, many eighteenth-century theories of art and beauty, and most of Kantian aesthetics, not to speak of many contemporary critical and philosophical approaches, namely: (a) Russian formalists, with their notion of "device" as the way in which the work of art elicits a particular type of perception; (b) Ingarden's attention to the reading process, his notion of the literary work as a skeleton or "schematized structure" to be completed by the reader, and his idea, clearly due to Husserl's influence, of the dialectics between the work as an invariant and the plurality of profiles through which it can be concretized by the interpreter; (c) the aesthetics of Mukarovsky; (d) Gadamer's hermeneutics; and (e) the early German sociology of literature (see Holub 1984, 2).

As for contemporary semiotic theories, they took the pragmatic moment into account from the beginning. Even without speaking of the central role played by interpretation and "unlimited semiosis" in C. S. Peirce's thought, it would be enough to remark that Charles Morris, in *Foundations of a Theory of Signs* (1938), reminded us that a reference to the role of the interpreter was always present in Greek and Latin rhetoric, in the communication theory of the Sophists, and in Aristotle, not to mention Augustine, for whom signs were characterized by their producing an idea in the mind of their receivers.

During the 1960s, many Italian semiotic approaches were influenced by sociological studies on the reception of mass media. In 1965 at a convention held in Perugia on the relationship between television and its audience, myself, Paolo Fabbri, and others insisted that it was not enough to study what a message says according to the code of its senders but also what it says according to the codes of its addressees. The idea of "aberrant decoding" proposed at that time was further elaborated in my *La struttura assente* (1968). Thus in the 1960s the problem of reception was posited (or reposited) by semiotics as a reaction against (1) the structuralistic idea that a textual object was something independent of its interpretations, and (2) the stiffness of many formal semantics flourishing in the Anglo-Saxon world, when the very meaning of a term or of a sentence was studied as independent of its context. Only later were dictionary-like semantics challenged by encyclopedia-like models that tried to introduce into the core of the semantic representation pragmatic elements also; and only recently have the cognitive sciences and the field of artificial intelligence decided that an encyclopedic model seems to be the most convenient way to represent meaning and to process texts (on this debate see Eco 1976, 1983).

In order to reach such an awareness it has been necessary for linguistics to move toward pragmatic phenomena, and in this sense the role of speech-act theory should not be underestimated. In the literary domain Wolfgang Iser (1978) was probably the first to acknowledge the convergence between the new linguistic perspectives and the literary theory of reception, devoting as he did an entire chapter of *Der Akt des Lesens* to the problems raised by J. L. Austin and John Searle (five years before the first organic attempt to elaborate a theory of literary discourse based upon speech-act theory; see Pratt 1977).

Thus what Jauss in 1969 was already announcing as a profound change in the paradigm of literary scholarship was in fact a change taking place in the semiotic paradigm in general—even though, as I said, this change was not a brand-new discovery but rather the complex concoction of different venerable approaches that had at various times characterized the history of aesthetics and a great part of the history of semiotics. Nevertheless it is not true that *nihil sub sole novi*. Old (theoretical) objects can reflect a different light under the sun's rays, according to the season.

I remember how outrageous it sounded to many when, in *Opera aperta* (1962), I stated that artistic and literary works, by forseeing a system of psychological, cultural, and historical expectations on the part of their addressees, try to produce what Joyce called an "ideal reader."[1]

Obviously, at that time, speaking of works of art, I was interested in the fact that such an ideal reader was obliged to suffer an ideal insomnia in order to question the book *ad infinitum*. If there is a consistent difference between *Opera aperta* (1962) and *The Role of the Reader* (1979), it is that in the latter book I try to find the roots of artistic "openness" in the very nature of any communicative process as well as in the very nature of any system of signification (as already advocated by my *A Theory of Semiotics* [1976]).

In any case, in 1962 my problem was how and to what extent a text should foresee the reactions of its addressee. In *Opera aperta*—at least at the time of the first Italian edition, written between 1957 and 1962—I was still moving in a presemiotic area, inspired as I was by information theory, the semantics of I. A. Richards, the epistemology of Jean Piaget, Maurice Merleau-Ponty's phenomenology of perception, transactional psychology, and the aesthetic theory of interpretation of Luigi

Pareyson. In that book, and with a jargon of which I now feel ashamed, I wrote that

> now we must shift our attention from the message, as a source of possible information, to the communicative relationship between message and addressee, where the interpretative decision of the receptor contributes in establishing the value of the possible information.
>
> If one wants to analyze the possibilities of a communicative structure one must take into account the receptor pole. To consider this psychological pole means to acknowledge the formal possibility—as such indispensable in order to explain both the structure and the effect of the message—by which a message signifies only insofar as it is interpreted from the point of view of a given situation—a psychological as well as a historical, social, and anthropological one.[2]

All these assumptions sounded pretty polemical in the 1960s because the structuralist orthodoxy was still operating under the standards of the "aesthetic-formalist" third paradigm designated by Jauss (1969). In 1967, speaking of my book *Opera aperta*, just translated into French, Claude Lévi-Strauss said in the course of an interview that he was reluctant to accept my perspective because a work of art

> is an object endowed with precise properties that must be analytically isolated, and this work can be entirely defined on the grounds of such properties. When Jakobson and myself tried to make a structural analysis of a Baudelaire sonnet, we did not approach it as an "open work" in which we could find everything that has been filled in by the following epochs; we approached it as an object which, once created, had the stiffness—so to speak—of a crystal; we confined ourselves to bringing into evidence these properties.

I have already discussed this opinion in the introductory chapter of my *The Role of the Reader*, making it clear that, by stressing the role of the interpretative choice in making up the sense of a text, I was not assuming that in an "open work" one can find that "everything" has been filled in by its different *empirical* readers, irrespective or in spite of the properties of the textual objects. I was, on the contrary, assuming that an artistic text contained, among its major analyzable properties, certain structural devices that encourage and elicit interpretative choices. However, I am quoting that old discussion in order to show how daring it was, during the 1960s, to introduce the interpretative moment, or if one wants, the act of reading, into the description and evaluation of the text to be read.

In *Opera aperta*, even though stressing the role of the interpreter ready to risk an ideal insomnia in order to pursue infinite interpretations, I was insisting that one ought always to question a text as an object, and not on the mere grounds of one's personal drives. Depending as I was on the aesthetics of interpretation of Luigi Pareyson, I was still speaking of a dialectics between fidelity and freedom. I am stressing this point because if during the "structural sixties" my addressee-oriented position (neither so provocative nor so unbearably original) appeared so "radical," today it would sound pretty conservative, at least from the point of view of the most radical reader-response theories.

2. A *Web of Critical Options*

The opposition between a generative approach (according to which the theory isolates the rules for the production of a textual object that can be understood independently of its effects) and an interpretative approach is not homogeneous with the triangular contrast, widely discussed in the course of a secular critical debate between interpretation as research into the *intentio auctoris*, interpretation as the research into the *intentio operis*, and interpretation as the imposition of the *intentio lectoris*.

The classical debate aimed at finding in a text either (*a*) what its author intended to say, or (*b*) what the text said independently of the intentions of its author. Only after accepting the second horn of the dilemma was the question whether to find in a text (1) what it says by virtue of its textual coherence and of an original underlying signification system, or (2) what the addressees find in it by virtue of their own systems of signification or their wishes and drives.

Such a debate is of paramount importance but its terms only partially overlap the opposition generation/interpretation. One can describe a text as generated according to certain rules without assuming that its author intentionally and consciously followed them. One can adopt a hermeneutic viewpoint leaving unprejudiced whether the interpretation must find what the author meant or what Being says through language— in the second case, leaving unprejudiced whether the voice of Being is influenced by the drives of the addressee or not. If one crosses the opposition generation/interpretation with the trichotomy of intentions one can get six potential different theories and critical methods.

Facing the possibility displayed by a text of eliciting infinite or indefinite interpretations, the Middle Ages and Renaissance reacted by embracing two different hermeneutic options. Medieval interpreters looked for a plurality of senses without refusing a sort of identity principle (a text cannot support contradictory interpretations), whereas the symbolists of the Renaissance, following the idea of the *coincidentia oppositorum*, defined the ideal text as the one that allows the most contradictory readings (see Eco 1985).

Moreover, the adoption of the Renaissance model generates a secondary contradiction, since a hermetico-symbolical reading can search in a text for either (1) the infinity of senses planned by the author, or (2) the infinity of senses that the author ignored. Naturally, option (2) generates a further choice, namely, whether these unforeseen senses are discovered because of the *intentio operis* or in spite of it, forced into the text by an arbitrary decision of the reader.

Even if one says, as Paul Valery did, that "il n'y a pas de vrai sens d'un texte," one has not yet decided on which of the three intentions the infinity of interpretations depends. Medieval and Renaissance Kabbalists maintained that the Torah was open to infinite interpretations because it could be rewritten in infinite ways by combining its letters, but such an infinity of readings (as well as of writings)—certainly dependent on the initiative of the reader—was nonetheless planned by the divine Author.

To privilege the initiative of the reader does not necessarily mean to guarantee the infinity of readings. If one privileges the initiative of the reader one must also consider the possibility of an active reader who decides to read a text univocally: it is a privilege of fundamentalists to read the Bible according to a single literal sense.

We can conceive of an aesthetics claiming that poetic texts can be infinitely interpreted because their author wanted them to be read this way; or an aesthetics that claims that texts must be read univocally in spite of the intentions of their authors, who are subject to the laws of language and once have written something are bound to read what they wrote in the only authorized and possible sense. One can read as infinitely interpretable a text conceived as absolutely univocal (see, for instance, the reading performed by Derrida upon a text of Searle in "Signature, evenement, contexte"), as well as one can perform psychedelic trips upon a text that cannot be but univocal according to the *intentio operis* (for instance, when one muses oneirically upon a railway timetable). Alternatively, one can read as univocal a text that its author wanted infinitely interpretable (this would be the case of fundamentalists if by chance the Kabbalists were right), or read univocally a text that from the point of view of linguistic rules should be considered rather ambiguous (for instance, reading *Oedipus rex* as a plain mystery story where what counts is only to find out the guilty one).

It is in the light of this embarrassingly vast typology that we should reconsider many contemporary critical currents that can superficially be ranked, all together, under the heading of response-oriented theories. For instance, classical sociology of literature records what readers do with a text and can remain basically uninterested in deciding on which intention what they do depends, since it simply describes social usages, socialized interpretations, and the actual public effect of texts, not the formal devices or the hermeneutic mechanisms that have produced them. The aesthetics of reception, on the other hand, maintains that a literary work is enriched by the various interpretations it undergoes along the centuries and, while considering the dialectics between textual devices and the readers' horizon of expectations, does not deny that every interpretation can and must be compared with the textual object and with the *intentio operis*. Likewise, semiotic theories of interpretative cooperation, like my theory of a Model Reader, look at the textual strategy as a system of instructions aimed at producing a possible reader whose profile is designed by and within the text and who can be extrapolated from it and described independently of and even before any empirical reading.

In a totally different way the most radical practices of deconstruction privilege the initiative of the reader and reduce the text to an ambiguous bunch of still-unshaped possibilities, thus transforming texts into mere stimuli for the interpretative drift.

3. An Apology of the Literal Sense

Every discourse on the freedom of interpretation must start from a defense of literal sense. In 1985 Ronald Reagan, during a microphone test before a public speech, said p (namely, "In a few minutes I'll push the red button and I'll start bombing the Soviet Union," or something similar). P was—as Linear Textual Manifestation—an English sentence that according to common codes means exactly what it intuitively means. If you prefer, once provided an intelligent machine with rules for paraphrasing, p could be translated as "the person uttering the pronoun 'I' will in the next approximately 200 seconds send American missiles toward the Soviet territory." If texts have intentions, p had the intention to say so.

The newsmen who heard p wondered whether its utterer too had the intention to say so. Asked about that, Reagan said that he was joking. He said so—as far as the *intentio operis* was concerned—but according to the *intentio auctoris* he only *pretended* to say so. According to common sense, those who believed that the sentence meaning coincided with the intended authorial meaning were wrong.

In criticizing Reagan's joke, some newsmen, however, tried to make an innuendo (*intentio lectoris*) and inferred that Reagan's real intention was to suggest nonchalantly that he was such a tough guy that, if he wanted, he could have done what he only pretended to do (also because he had the performative power of doing things with words).

This story is scarcely suitable for my purposes because it is a report about a fact, that is, about a "real" communicative interchange during which senders and addressees had the chance of checking the discrepancies between sentence meaning and authorial meaning. Let us suppose, then, that this is not a story about a fact but a pure story (told in the form "once a man said so and so, and people believed so and so, and then that man added so and so . . ."). In this case we have lost any guarantee about authorial intention, this author having simply become one of the characters of the narration. How to interpret this story? It can be the story of a man making a joke, the story of a man who jokes but shouldn't, the story of a man who pretends to joke but as a matter of fact is uttering a menace, the story of a tragic world where even innocent jokes can be taken seriously, the story of how the same jocular sentence can change in meaning according to the status and the role of its utterer. . . . Would we say that this story has a single sense, or that it has all the listed ones, or that only some of them can be considered the "correct" ones?

Two years ago Derrida wrote me a letter informing me that he and others were establishing in Paris a Collège International de Philosophie and asking me for a letter of support. I bet that Derrida was assuming that:

—I had to assume that he was telling the truth;

—I had to read his program as a univocal discourse as far as both the actual situation and his projects were concerned;

—my signature requested at the end of my letter would have been taken more seriously than Searle's at the end of "Signature, evenement, contexte."

Naturally, according to my *Erwartungshorizon* Derrida's letter could have assumed for me many other additional meanings, even the most contradictory ones, and could have elicited many additional inferences about its "intended meaning"; nevertheless, any additional inference ought to be based upon its first layer of allegedly literal meaning. I think that Derrida could not but agree with me: in *Of Grammatology* he reminds his readers that "without [all the instruments of traditional criticism] . . . critical production will risk developing in any direction at all and authorize itself to say almost anything. But this indispensable guardrail has always only *protected*, it has never *opened* a reading" (Derrida 1976, 158). I feel sympathetic with the project of opening readings but I also feel the fundamental duty of protecting them in order to open them, since I consider it risky to open in order to protect. Thus, coming back to Reagan's story, my conclusion is that, in order to extrapolate from it any possible sense one is first of all obliged to recognize that it had a literal sense, namely, that on a given day a man said *p* and that *p*, according to the English code, means what it intuitively means.

4. Two Levels of Interpretation

Before going ahead with the problem of interpretation we must first settle a terminological question. We must distinguish between *semantic* and *critical* interpretation (or, if one prefers, between *semiosic* and *semiotic* interpretation). Semantic interpretation is the result of the process by which an addressee, facing a textual linear manifestation, fills it up with a given meaning. Every response-oriented approach deals first of all with this type of interpretation, which is a natural semiosic phenomenon. Critical interpretation, on the other hand, is a metalinguistic activity—a semiotic approach—which aims at describing and explaining the formal reasons for which a given text produces a given response (and in this sense it can also assume the form of an aesthetic analysis).

In this sense every text is susceptible to both semantic and critical interpretation, but only few texts consciously foresee both kinds of response. Ordinary sentences (like "give me that bottle" or "the cat is on the mat" uttered by a laymen) expect only a semantic response. But aesthetic texts or sentences like "the cat is on the mat" uttered by a linguist as an example of possible semantic ambiguity also foresee a critical interpreter. Likewise when I say that every text designs its *own* Model Reader, I am in fact implying that many texts aim at producing *two* Model Readers, a first-level or a naive one, supposed to understand semantically what the text says, and a second-level or critical one, supposed to appreciate the way in which the text says so. A sentence like "they are flying planes" foresees a naive reader who keeps wondering which meaning to choose—and who supposedly looks at the textual environment or at the circumstance of utterance in order to support the best choice—and a critical reader able to explain univocally and formally the syntactic reasons that make the sentence ambiguous. Similarly, a mystery tale displays an astute narrative strategy in order to produce a naive Model Reader eager to fall into the traps of the narrator (to feel fear or to suspect the innocent one), but usually wants also to produce a critical Model Reader able to enjoy, at a second reading, the brilliant narrative strategy by which the first-level naive reader has been designed.[3]

In the light of the above distinctions let me now discuss a distinction between two interpretative theories of our time proposed by Richard Rorty in his essay "Idealism and Textualism" (in *Consequences of Pragmatism* [1982]). Rorty says that in the present century "there are people who write as if there were nothing but texts" and makes a distinction between the two kinds of textualism. The first is instantiated by those who disregard the intention of the author and look in the text for a principle of internal coherence and/or for a sufficient cause for certain very precise effects it has on a presumed ideal reader. The second is instantiated by those critics who consider every reading a misreading (the "misreaders"). For them, says Rorty, "the critic asks neither the author nor the text about their intentions but simply beats the text into a shape which will serve his own purpose. He makes the text refer to whatever is relevant to that purpose." In this sense their model "is not the curious collector of clever gadgets taking them apart to see what makes them work and carefully ignoring any extrinsic end they may have, but the psychoanalyst blithely interpreting a dream or a joke as a symptom of homicidal mania."

Rorty thinks that both positions are a form of pragmatism (pragmatism being for him the refusal to think of truth as correspondence to reality—and reality being, I assume, both the external referent of the text and the intention of its author) and suggests that the first type of theorist is a weak pragmatist because "he thinks that there really is a secret and that once it's discovered we shall have gotten the text right," so that for him "criticism is discovery rather than creation." The strong pragmatist, on the other hand, does not make any difference between finding and making.

I can accept such a characterization, but with two emendations.

First of all, in what sense does a weak pragmatist, when trying to find the secret of a text, aim at getting this text right? One has to decide if by "getting the text right" one means a right semantic or a right critical interpretation. Those readers who, according to the Jamesian metaphor proposed by Iser (1978, chap. 1), look into a text in order to find in it "the figure in the carpet," a single unrevealed secret meaning, are—I think—looking for a sort of "concealed" semantic interpretation. But the critic looking for the "secret code" probably looks *critically* for the describable strategy that produces infinite ways to get a text semantically right. To analyze and describe the textual "devices" of *Ulysses* means to show how Joyce acted in order to create many alternative figures in his carpet, without deciding how many they can be and which of them are the best ones. Obviously, since—as I shall tell later—even a critical reading is always conjectural, there can be many ways of finding out that secret code, but to look for it does not mean that one wants to reduce a text to a univocal semantic reading. Thus I do not think that the first type of textualist designated by Rorty is necessarily a "weak" pragmatist.

Second, I suspect that many "strong" pragmatists are not pragmatists at all, at least in Rorty's sense, because the "misreader" employs a text in order to know something that stands outside the text—and that is in some way more "real" than the text itself, namely, the unconscious mechanism of *la chaîne signifiante*. In any case, even though a pragmatist, certainly the misreader is not a "textualist." Probably misreaders think, as Rorty assumes, that there is nothing but texts; however, they are interested in every possible text except the one they are reading. As a matter of fact "strong" pragmatists are concerned only with the infinite semantic readings of the text they are beating, but I suspect that they are scarcely interested in the way it works.

5. Interpretation and Use

I can accept the distinction proposed by Rorty, but I see it as a convenient opposition between *interpreting* (critically) and merely *using* a text. To critically interpret a text means to read it in order to discover, along with my reactions to it, something about its nature. To use a text is to start from a stimulus in order to get something else, even accepting the risk of misinterpreting it from the semantic point of view. If I pull out the pages of my Bible to wrap my pipe tobacco in, I am using this Bible, but it would be daring to call me a textualist—even though I am, if not a strong pragmatist, certainly a very pragmatic person. If I get sexual enjoyment from a pornographic book, I am not using it, because in order to elaborate my sexual fantasies I had to semantically interpret its sentences. On the other hand, if—let us suppose—I look into the *Elements* of Euclid to infer that their author was a scopophiliac, obsessed with abstract images, then I am using it, because I renounce interpreting its definitions and theorems semantically.

The quasipsychoanalytic reading that Derrida gives of Poe's *Purloined Letter* in *Le facteur de la vérité* represents a good critical interpretation of that story. Derrida insists that he is not analyzing the unconscious of the author but rather the unconscious of the text. He is interpreting because he respects the *intentio operis*. When he draws an interpretation from the fact that the letter is found in a paper holder hanging from a nail under the center of a fireplace, he first takes "literally" the possible world designated by the narration as well as the sense of the words used by Poe to stage this world. Then he tries to isolate a second "symbolic" meaning that this text is conveying, probably beyond the intentions of the author. Right or wrong, Derrida supports his second-level semantic interpretation with textual evidence. In doing so he also performs a critical interpretation, because he shows how the text can produce that second-level semantic meaning.

For the opposite approach, let us consider Marie Bonaparte's method of analyzing Poe's work. Part of her reading represents a good example of interpretation. For instance, she reads "Morella," "Ligeia," and "Eleonora" and shows that all three texts have the same underlying "fabula": a man in love with an exceptional woman who dies of consumption swears eternal grief; but he does not keep his promise and loves another woman; finally the dead one reappears and wraps the new one in the mantle of her funereal power. In a nontechnical way Marie Bonaparte identifies in these three texts the same actantial structures, speaks of the structure of an obsession, but reads that obsession as a textual one and in so doing reveals the *intentio operis*. Unfortunately such a beautiful textual analysis is interwoven with biographical remarks that connect textual evidences with aspects (known from extratextual sources) of Poe's private life. When she says that Poe was dominated by the impression he felt as a child when he saw his mother, dead of consumption, lying on the catafalque, when she says that in his adult life and in his work he was morbidly attracted to women with funereal features, when she reads his stories populated by living corpses in order to explain his personal necrophilia—then she is using and not interpreting texts.

6. Interpretation and Conjecture

It is clear that I am trying to keep a dialectical link between *intentio operis* and *intentio lectoris*. The problem is that, if one perhaps knows what is meant by "intention of the reader," it seems more difficult to define abstractly what is meant by "intention of the text."

The intention of the text is not displayed by the Textual Linear Manifestation. Or, if it is so displayed, it is in the sense of the purloined letter. One has to decide to "see" it. Thus it is possible to speak of the intention of the text only as the result of a conjecture on the part of the reader. The initiative of the reader basically consists in making a conjecture about the intention of the text.

A text is a device conceived in order to produce its Model Reader. I repeat that this Reader is not the one who makes the *only right* conjecture. A text can foresee a Model Reader entitled to try infinite conjectures. The empirical reader is only an actor who makes conjectures about the kind of Model Reader postulated by the text. Since the intention of the text is basically to produce a Model Reader able to make conjectures about it, the initiative of the Model Reader consists in figuring out a Model Author who is not the empirical one and who, in the end, coincides with the intention of the text.

Thus, more than a parameter to use in order to validate the

interpretation, the text is an object that the interpretation builds up in the course of the circular effort of validating itself on the basis of what it makes up as its result. I am not ashamed to admit that I am so defining the still valid hermeneutic circle.

The logic of interpreting is the Peircian logic of *abduction* (see Eco and Sebeok 1983). To make a conjecture means to figure out a Law that can explain a Result. The "secret code" of a text is such a Law. One could say that in the natural sciences the conjecture has to test only the Law since the Result is under the eyes of everybody, whereas in textual interpretation only the discovery of a "good" Law makes the Result acceptable. But I do not think that the difference is so clear-cut. Even in the natural sciences no fact can be taken as a significant Result without someone having first and vaguely decided that this fact among innumerable others can be selected as a curious Result to be explained. To isolate a fact as a curious Result means to have already obscurely thought of a Law of which that fact could be the Result. When I start reading a text I never know from the beginning if I am approaching it from the point of view of a suitable intention. My initiative starts becoming exciting when I discover that my intention could meet the intention of the text.

How to prove a conjecture about the *intentio operis*? The only way is to check it against the text as a coherent whole. This idea, too, is an old one and comes from Augustine (*De doctrina christiana*): any interpretation given of a certain portion of a text can be accepted if it is confirmed and must be rejected if it is challenged by another portion of the same text. In this sense internal textual coherence controls the otherwise uncontrollable drives of the reader.

Once Borges suggested that it would be exciting to read the *Imitation of Christ* as if it was written by Celine. The game is amusing and could be intellectually fruitful. With certain texts it could suggest new and interesting interpretations. It cannot, however, work with Thomas à Kempis. I tried: I discovered sentences that could have been written by Celine ("Grace loves low things and is not disgusted by thorny ones, and likes filthy clothes . . . .") But this kind of reading offers a suitable "grid" for very few sentences of the *Imitatio*. All the rest, most of the book, is resistant to this reading. If, on the contrary, I read the book according to the Christian medieval encyclopedia, it appears textually coherent in each of its parts.

Besides, no responsible deconstructionist has ever challenged such a position. Hillis Miller (in "On Edge") said that "the readings of deconstructive criticism are not the willful imposition by a subjectivity of a theory on the texts, but are coerced by the texts themselves" (p. 611). Elsewhere (in *Thomas Hardy: Distance and Desire*) he writes that "it is not true that . . . all readings are equally valid. . . . Some readings are certainly wrong. . . . To reveal one aspect of a work of an author often means ignoring or shading other aspects. . . . Some approaches reach more deeply into the structure of the text than others."

7. The falsifiability of misinterpretations (Fälschungmöglichkeit)

We can thus accept a sort of Popper-like principle according to which if there are no rules that help to ascertain which interpretations are the "best ones," there is at least a rule for ascertaining which ones are "bad." As I said above, this rule says that the internal coherence of a text must be taken as parameter for its interpretations. But in order to do so one needs, at least for a short time, a metalanguage that permits the comparison between a given text and its semantic or critical interpretations. Since any new interpretation enriches the text and the text consists in its objective Linear Manifestation plus the interpretations it received in the course of history, this metalanguage should also allow the comparison between a new interpretation and the old ones.

I understand that from the point of view of a radical deconstruction theory such an assumption can sound unpleasantly neopositivistic, and that Derrida's notion of deconstruction and drift challenges the very possibility of a metalanguage. But a metalanguage does not have to be different from (and more powerful than) ordinary language. The idea of interpretation requires that a "piece" of ordinary language be used as the *interpretant* (in the Peircian sense) of another "piece" of ordinary language. When one says that *man* means "human male adult" one is interpreting ordinary language through ordinary language, and the second "sign"—said Peirce—is the interpretant of the first one, as the first can become the interpretant of the second.

The metalanguage of interpretation is not different from its object language. It is a portion of the same language and in this sense interpretation is a function that every language performs when it speaks of itself. It is not a matter of asking if this can be done. We are doing it, everyday.

The provocative self-evidence of my last argument suggests that we can prove it only by showing that any of its alternatives is self-contradictory.

Let us suppose that there is a theory that *literally* (not metaphorically) asserts that every interpretation is a misinterpretation.

Let us suppose that there are two texts α and β and that *one* of them (we know which one) has been proposed to a reader in order to elicit the textually recorded misinterpretation Σ.

Take a literate subject X, previously informed that any interpretation must be a misinterpretaton, and give him/her the three texts, α, β, and Σ.

Ask X if Σ misinterprets α or β.

Supposing that X says that Σ is a misinterpretation of α, would we say that X is right? Supposing on the contrary that X says that Σ is a misinterpretation of β, would we say that X is wrong?

In both cases, to approve or disaprove X's answer means to believe not only that a text controls and selects its own interpretations but also that it controls and selects its own misinterpretations. The one approving or disproving X's answers would then act as one who does not really believe that every interpretation is a misinterpretation, since he/she would use the original text as a parameter for discriminating between texts that misinterpret it and texts that misinterpret something else. This move would presuppose a previous interpretation of α which should be considered the only correct one, as well as a metalanguage that describes A and shows on which grounds Σ is or is not a misinterpretation of it.

It would be embarrassing to maintain that a text elicits only misinterpretations except when it is correctly interpreted by the warrant of the misinterpretations of other readers. But this is exactly what happens with a radical theory of misinterpretation.

There is another way to escape the contradition. One should assume that every answer of X is the good one. Σ can be indifferently the misinterpretation of α, of β, and of any other possible text. But at this point why define Σ (which is undoubtedly a text in its own right) as the misinterpretation of something else? If it is the misinterpretation of everything, it is, then, the misinterpretation of nothing. It exists for its own sake and does not need to be compared with any other text.

The solution is elegant but a little inconvenient. It destroys definitely the very category of textual interpretation. There are texts, but of these nobody can speak. Or, if one speaks, nobody can say what one says. Texts, at most, are used as stimuli to produce other texts, but once a new text is produced, it cannot be referred to its stimulus.

8. Conclusions

To defend, as I have done, the rights of interpretation against the mere use of a text does not mean that texts must never be used. We are using texts everyday and we need to do so, for many respectable reasons. It is only important to distinguish use from interpretation.

A critical reader could also say why certain texts have been used in a certain way, finding in their structure the reasons of their use or misuse. In this sense a sociological analysis of the free uses of texts can support a further interpretation of them.

In any case use and interpretation are abstract theoretical possibilities. Every empirical reading is always an unpredictable mixture of both. It can happen that a play started as use ends by producing a fruitful new interpretation—or vice versa. Sometimes to use texts means to free them from previous interpretations, to discover new aspects of them, to realize that they have previously been illicitly interpreted, to find out a new and more explicative *intentio operis*, to realize that too many uncontrolled intentions of readers (perhaps disguised as a faithful quest for the intention of the author) have polluted and obscured them.

There is also a *pretextual* reading, performed not in order to interpret the text but to show how much language can produce unlimited semiosis. Such a pretextual reading has a philosophical function, and many of the examples of deconstruction provided by Derrida belong to this kind of activity. It has happened that a legimate philosophical practice has been taken as a model for literary criticism and for a new trend in textual inter-pretation.

Our theoretical duty was to acknowledge that this happened and to show why it should not have happened.

GIANNI VATTIMO
Metaphysics, Violence, Secularization

Contemporary philosophy has become ever more aware that the reasons for the distrust of metaphysics and the program of "going beyond" it, variously formulated in the nineteenth and twentieth centuries, are ethical rather than "theoretical." This is probably, and perhaps paradoxically (since we are dealing with the author of a book like *Beyond Good and Evil*), the most characteristic effect of the Nietzsche renaissance in European thought of the past twenty years. In Nietzsche's philosophy there is, on the one hand, a sort of *summa* of the "unmaskings" of metaphysics that philosophy has proposed in the past century or so, from the Marxian critique of ideology to the Freudian discovery of the unconscious, by way of positivism and the birth of the "human sciences." But there is also, as more characteristic of and specific to his thought, a radical "unmasking of unmasking" according to which even the idea of a truth that reveals a masking, of the attempt and claim to reach a solid "ground" beyond ideologies and every form of false consciousness, is, precisely, still a "human, all too human" devotion, still a mask. If, as Nietzsche maintains, we must distrust metaphysics—that is to say, for him as for us, the belief in a stable structure of Being that governs becoming and gives meaning to knowledge and norms to conduct—it is finally not "on strict grounds of knowledge" (as Nietzsche writes in an aphorism entitled "An Affectation in Parting," no. 82 of *The Wanderer and His Shadow*).[1] If that were so, we would remain always still prisoner of another metaphysics, of a theory that opposes a "true truth" to the errors to be unmasked, thereby perpetuating the game from which we wish to escape in getting out of metaphysics. Of course, even saying that we "want to" (or must, or cannot not, etc.) escape the game and the mechanism of metaphysics implies that one speaks in the name of a subject, a "we" that seems already completely internal to that metaphysical logic from which one wishes to escape. Nietzsche is well aware of this. Perhaps that is why he insists so much, in his late writings, on the opposition between "overman" and "slaves," or, as Pierre Klossowski notes, represents his own doctrine in the form of a "plot," for both are ways to exclude oneself explicitly from the horizon of "universal" affirmations characteristic of traditional metaphysical philosophy.[2] Even and above all the most comprehensive and scandalous proposition of his texts, the statement "God is dead," is not a metaphysical thesis argued and demonstrated to the ideal "we" of human reason. It is the tale of an experience, an appeal to others that they might discover it in themselves, constituting, on this basis, a "we" to whom and in whose name Nietzsche might speak.

Does this "we" that constitutes itself a posteriori—beginning with the community of a recognized experience, from the reception of an appeal—truly escape from the unmasking of metaphysics and its universal structures? Or does the difficulty of unmasking metaphysics and the very notion of a truth not spoken always in the name of a "we" show, at least to some extent, that in the final analysis one cannot really get out of metaphysics? And again, can we dissolve this problem by admitting that the metaphysical "we" is really just a rule of the literary genre that we choose to adopt when we decide (but to what extent in a purely arbitrary way?) to do philosophy? One must probably give affirmative answers to all three of these questions, and thus follow a sort of widespread consensus, implicit or explicit, in contemporary thought. We accept the idea that the "we" of philosophical discourse is not ideally given in the realm of a universal and eternal reason but is constituted historically as the generalizability of experience (and hence also dependent upon the constitution of a society in which communication tends toward universality and there is something like "public opinion"). Second, this recognition does not suddenly place us in a horizon different from that of metaphysics, since in fact metaphysics has worked exactly in this way, even if with a different awareness, in constituting the "we" it believed it found given as human essence. Third, it is quite true that we therefore continue to belong to metaphysics, but what binds us to it is "only" the continuity—if not casual and arbitrary, certainly historically contingent—of a "literary genre" and of the culture to whose constitution it contributes.

To dwell upon the difficulties that arise when, wishing to call metaphysics into question, one continues to use its language—by adopting, for example, the "fiction" of a universal subject like the "we" of traditional philosophical statements—is not a way to guarantee the validity, logical correctness, and truth of conclusions at which one wants to arrive, taking care that the initial steps, on which all the rest will depend, are "grounded." Indeed, it is metaphysics that takes care to "ground" rigorously evey step. Aware that the circle is not necessarily a vicious one but on the contrary guarantees the only possible access to our topic,[3] we take on this problem as well, almost by accumulation, among the dimensions, stratifications, and ambiguities that burden the discourse of metaphysics and its "going beyond" in contemporary philosophy, in the philosophy that has learned from Nietzsche.

Having "listened to" Nietzsche, then, we do not want to, we cannot, we must not (any longer) limit ourselves to unmasking that inherited metaphysics in the name of a truer ground. Nietzsche has taught us to distrust the very idea of a true ground. Even words like "suspicion" and "distrust" that recur so often in his texts are terms that indicate attitudes more practical than theoretical. This teaching, and the underlying experience that Nietzsche was the first to live and formulate in radical fashion, are no longer only Nietzsche's. What he theorized—that metaphysics is only a form of the will to power—has widely penetrated twentieth-century thought with sometimes differing but always closely related meanings. The critique of metaphysics that most endured in his thought (and that thus determines our point of departure) is the critique that would unmask metaphysics as a manifestation of violence. The refusal of metaphysics inspired by reasons "of knowledge," like that exemplified in Rudolf Carnap's famous text against Heidegger in 1932, no longer has any force.[4] Not only has this text and the attitude it expresses been forgotten, but it no longer enjoys any visible reception even in neo-empiricist thought. In fact, the paradoxical fortune that the term metaphysics enjoys precisely in authors and contexts of an "analytic"[5] tendency—for example, in Karl Popper—may be considered its "liquidation." If there is a "friendly" attitude toward metaphysics today, it can be found not only in proponents of a neoclassical thought, but also in philosophers of neo-empiricist background as well. A not-so-paradoxical fact, if one remembers the close link that Heidegger on good grounds points out between the inheritance of metaphysics and modern "scientism."

In any case, it is not in the name of theoretical motivations that one speaks today of going beyond metaphysics, as though its fault were that of furnishing a distorted and false knowledge of a reality that could be understood adequately by other forms of knowledge, above all by science. Those who distrust metaphysics and think of going beyond it move instead within the horizon defined by Nietzsche when he writes that metaphysics is "an attempt to take by force the most fertile fields."[6] And Heidegger, the most radical theorist of the necessity of going beyond metaphysics, also does not have theoretical motivations. *Being and Time* takes its point of departure instead from the impossibility of thinking the existence of man within the categories of traditional metaphysics, above all in the light of the notion of Being as disclosed presence. This impossibility does not, however, constitute a problem in the sense that a more "adequate" notion of Being, capable of being applied correctly to the Being of man as well, would be necessary. *Being and Time*, in fact, already contains Heidegger's critique of the notion of truth as adequation of proposition to thing. We cannot presume, then, that traditional metaphysics (still called ontology) deserves the "destruction" that Heidegger proposes for it (see par. 6) because it does not satisfy a criterion of truth as correspondence. All we can do is read as implicit in *Being and Time* what will become ever clearer in the later Heidegger: we are called to go beyond metaphysics "because" it discloses itself today in *Ge-Stell*, in the world of total technico-scientific organization in which being-there not only does not allow itself to "be thought" (as it appeared in *Being and Time*) on account of the domination of simple presence, but also cannot "be there" in a radical sense. Heidegger's references to the atomic bomb and the desertification of the world in talks and addresses of the 1950s and later are not merely "occasional," dictated by the good intention of joining his voice to those worried about the future of the human race in the epoch of great technology of destruction. They contain the "essence" of his thought insofar as all the effort of that thought, beginning with *Being and Time*, to "recollect" Being by going beyond metaphysics is motivated by the experience of violence. That this is not so evident in his texts is partly a result of self-misunderstandings of the sort that lie at the origin of his support for Nazism. More important for us, it is also because Heidegger, like Nietzsche (and for the same reasons, linked to the unmasking of unmasking), does not oppose another ethical metaphysics of nonviolence to metaphysical violence. He is therefore not in a position to propose clear-cut denunciations but must follow the path of *Verwindung*, of acceptance-distortion, that leads out of metaphysics only by way of its secularizing continuation.

The necessity of following this path seems to emerge from the same difficulties, genuine aporias, that the most powerful denunciations of the violent essence of metaphysics—those of Theodor Adorno and Emmanuel Lévinas—also run up against.

In the last chapter of *Negative Dialectics*, "Meditations on Metaphysics," Adorno not only does not see metaphysics and modernization as alternatives, seeing them instead as closely connected (metaphysics is rationalization which, in modern society, becomes the actualized technico-scientific organization of society), but also, at least to some extent, seems to consider the extreme explosion of violence an unveiling and decisive step along the way of a possible "going beyond" of the metaphysics that inspires that violence and is disclosed in it.[7] Using a language different from Adorno's, I would say that it is perhaps only after Auschwitz that Being can give itself in its authentic minimal and micrological essence (though without, obviously, attributing any necessary providentiality to Auschwitz).

The crematory ovens at Auschwitz, Adorno says, are not only a consequence of a certain rationalistic vision of the world, but also and above all an anticipatory image of what the administered world is and does in the normal course of events that affirms and renders universal "the indifference of each individual life" (ND, 362). The atrocity of Auschwitz makes it impossible still to affirm "that the immutable is truth, and that the mobile, transitory is appearance," and not only insofar as "our metaphysical faculty is paralyzed because actual events have shattered the basis on which speculative metaphysical thought could be reconciled with experience" (ND, 361, 362). Metaphysics turns out to be discredited also and above all because the indifference to each individual life, to the rights of the contingent and transitory, is what has always constituted its essential content. In some way, Auschwitz merely emphasizes all this by disclosing intolerable violence. (The negativity of Adorno's dialectics, which, as we will see later, places the *telos* of reconciliation in an unreachable future, dissolves for him as well the possibility of a true philosophy of history. It is therefore difficult to say whether "after Auschwitz" we can expect some turning point in the history of violence and metaphysics that would be different and specific in respect to other moments in history, or whether Auschwitz is, for Adorno, only the most "extreme" manifestation—because more visible to us and more atrocious than other wars, massacres, or natural catastrophes—that we can experience or remember of a violent "structure" of existence in which, because reconciliation can only be utopia, there are no caesuras, stages, or significant changes. One can probably not decide such a question on the basis of Adorno's texts. It seems to me that Adorno's discourse would have to attribute to "after Auschwitz" the significance of a turning point, and not merely a value as rhetorical *exemplum* of violence linked to the human condition, in order to have a meaning that is not purely "edifying" and metaphysico-descriptive.)

In any case, it is because metaphysics, and culture in general, covers and forgets the rights of the living immediate that it prepares Auschwitz: "That we no longer know what we used to feel before the dogcatcher's van, is both the triumph and failure of culture" (ND, 366) and of metaphysics.[8] The course of history "forces materialism upon metaphysics, traditionally the direct antithesis of materialism" (ND, 365), in the sense that it is the transitory, the repressed, the low, that now legitimately claims the rights of the essential. But it is not simply a question of "overturning Platonism": "Whoever pleads for the maintenance of this radically culpable and shabby culture becomes its accomplice, while the man who says no to culture is directly furthering the barbarism" (ND, 367). In some sense, it is literally very true that materialism presents itself as metaphysics: even to claim the rights of the transitory, the immediate, the low is to establish a "doubling," a form of transcendence of what should be with respect to what is: "What is a metaphysical experience? If we disdain projecting it upon allegedly primal religious experiences, we are most likely to visualize it as Proust did, in the happiness, for instance, that is promised by village names like Otterbach, Watterbach, Reuenthal, Monbrunn" (ND, 373).[9]

Not only is it metaphysical violence to cover and cancel the rights of the sensuous and transitory by affirming universal and abstract essences, it is equally violent and fetishistic to strip the sensuous of its dimension of alterity, negating that "promesse de bonheur" that the living insist on reading into it as illusory consolation and, in the end, as masked violence. The pure unmasking of the violence of the metaphysical universal would then be transformed into a nihilistic metaphysics, equally as violent as what it wishes to negate. Against this repetition of metaphysics, the sensuous and transitory would reassert its own rights: "And yet the lighting up of an eye, indeed the feeble tail-wagging of a dog one gave a tidbit it promptly forgets, would make the ideal of nothingness evaporate . . ." (ND, 380). In fact, the greatness of Kant for Adorno was not that of having placed reason within its limits by decreeing the impossibility of *knowing* the traditional objects of metaphysics: God, freedom, immortality, the totality of the world. Rather, it lay in having defined a "space" for those metaphysical ideas, a space that in the end is reduced to the appearance of the aesthetic but is no less true, even if it is only the problematic truth of the "promesse de bonheur" that Adorno takes pains to define with the (problematic) concept of a negative dialectics. The "true" ground of things, to be opposed to the violence of metaphysical universality, is not the total senselessness of nihilism—what Nietzsche would call reactive nihilism: "As in Kafka's writings, the disturbed and damaged course of the world is incommensurable also with the sense of its sheer senselessness and blindness; we cannot stringently construe it according to their principle. It resists all attempts of a desperate consciousness to posit despair as an absolute" (ND, 403–4).

The essence of the current post-Nietzschean, post-Heideggerian critique of metaphysics is here clearly defined, even if Adorno's answers are not entirely satisfactory. Thought revolts against metaphysics in the name of the historical experience of violence that appears to be linked to metaphysics—not only the violence of Auschwitz but also that of the society of total organization prepared and made possible by metaphysical essentialism and by all the procedures of repression of the transitory that constitute "culture." The offended consciousness reacts to Auschwitz by unmasking the claims of metaphysics. But no matter how contradictory this unmasking may appear, it is still a reaffirmation of *rights* which, as rights, are not pure, immediate facticity. Precisely that which is most ephemeral and uncertain in the sensuous world (the "lighting up of a eye," the "tail-wagging of a dog") carries within itself a promise of happiness that will not be silenced by an "unmasking" critique. The very reasons that inspire the critique of metaphysics also motivate a paradoxical recuperation of metaphysics—for Adorno, negative dialectics—located in "the no man's land between the border posts of being and nothingness" (ND, 381), the land that hosts the promise of happiness that shines forth in the dog who wags his tail. We are still within a metaphysical horizon characterized by the constant reappearance of the "ontological argument": "The concept is not real, as the ontological argument would have it, but there would be no conceiving it if we were not urged to conceive it by something in the matter" (ND, 404).

Adorno outlines this "recuperation" of metaphysics as negative dialectics through his insistence upon the Kantian category of *appearance* (both in *Negative Dialectics*, to which we here refer, and in *Aesthetic Theory*[10]) and upon that of *micrology* with its Benjaminian ring. These categories are rather precariously "grounded" in Adorno's discourse. It is difficult to see how the postulates of Kantian practical reason can be brought back to aesthetic experience without residues, as Adorno seems to intend. And as for "micrology," Negative Dialectics gets to the bottom of it only by means of an acrobatic leap, by evoking a notation familiar to contemporary music, the "presque rien" (ND, 407), as the most likely way to survive, to maintain oneself in the no-man's-land between being and nothingness. There is no real argument demonstrating the "why and wherefore" of the categories of appearance, the transitory, and micrology. Unless, of course, one consideres Adorno's dialectical mode of thought to be that argument. But it is precisely in the notion of negative dialectics that all the problems caused by these other notions are concentrated rather than dissolved. Adorno defines negative dialectics as "the very negation of negation that will not become a positing" (ND, 406), a knowledge "absolute" insofar as true in its unresolved problematicity.

The "promesse de bonheur" is essentially appearance, it *must* remain an unfulfilled promise. Opposed to the violent cancellation and repression of the sensuous and transitory performed by metaphysics in its traditional form is the right of the transitory as an expectation of happiness that would be a reconciliation of sensuous and intelligible, spirit and matter, the self and nature—in short, the beauty of Hegelian aesthetics or the generic essence of the Marxian unalienated man. This reconciled condition must remain always a promise because its realization would imply precisely the suppression of that dimension of alterity, nostalgia, aspiration, and openness that constitutes the right and truth of the transitory. Yet what makes the notion of negative dialectics problematic and precarious is not so much the constitutive unrealizability of the promised "bonheur" (an unrealizability that has taken leave of the still-religious background of Kantian postulates in order to bring everything back to the aesthetic domain of appearance), but rather the model of thought that continues to act as a base for such dialectics. In spite of all the sincere emphasis given appearance and micrology, Adorno's "bonheur" is still always thought according to the most classical metaphysical mechanisms of grounding. Going beyond metaphysics and its violence thus becomes a sort of exorcism, a movement to displace far away on a utopian horizon the feared and desired moment of access to grounding. Dialectical reconciliation is the exclusion of all transcendence, but this only constitutes the right of the transitory and sensuous against the "normalizing" claims of metaphysics. In other passages of the same text, metaphysical transcendence is explicitly brought back to the "separation of body and soul, as reflex of the division of labor" (ND, 400), and it would be difficult to find another "justification" in Adorno. And yet he must consider transcendence not only as an expression of alienation, but also as the source of the right of the sensuous in its necessary referral to something else. . . . In short, in a dialetical perspective, every tension between being and "ought to be," between the factual and the intellible worlds, is an expression of a split that must be reconciled: the tension is temporary and must be suppressed. In his concern to suspend indefinitely reconciliation—the moment in which dialectics becomes positing—Adorno seems to realize that what constitutes the violence of metaphysics is not so much the mechanism of transcendence, the referral to another order of reality that devalues and lowers that which is immediately given, but rather the mechanism of grounding, the process that claims to reach that promised "other" and to establish itself in its disclosed presence, in its *energeia*.

The difficulties that emerge from the notion of negative dialectics, then, seem only to express a more serious and more radical problematic encountered by every attempt to go beyond metaphysics without abandoning the conception of Being as disclosed presence that determined the development of metaphysics and that still dominates in the thought of Hegel and Marx to which Adorno remains tied.

This uncertainty about the true polemical objective of going beyond metaphysics (transcendence of the sensuous in the direction of a stable order or of a realm of essences? or rather the very idea that true Being is *Grund*, grounding, peremptory alterity that cannot be gone beyond, fully disclosed presence?) confers a particular fragility upon Adorno's negative dialectics. Adorno, of course, could defend negative dialectics in the name of micrology and the "presque rien" as thought that evades metaphysics in its very form by refusing, strictly speaking, the logic of grounding. In order to develop such a defense, however, he would have to articulate better the implications of his micrology which remain implicit, while the dominant model remains that of dialectical reconciliation—even if transferred to a utopian future—that is to say, the idea of true Being as presence. Although the identification of metaphysics as violence places Adorno on the horizon opened by Nietzsche and Heidegger, that he remains tied to dialectics and to the idea of Being it implies makes of him a still pre-Nietzschean and pre-Heideggerian thinker.

One can liquidate metaphysics as violence and rediscover metaphysics, in a different sense, as vindication of the right of the living to a transcendental dimension (the "promesse de bonheur") without radically criticizing the mechanism that has always, in its entire history, constituted the base of metaphysical violence: the mechanism of referral to *Grund*, of grounding. This is not *only* a reductive mechanism, as is often thought by those who read in too simplistic a way the Heideggerian definition of metaphysics as thought that forgets Being in favor of the entity. Together with the forgetfulness of the ontological difference, masking and forgetting of this very forgetfulness are constitutive of metaphysics for Heidegger. Such forgetting reveals itself precisely as the preservation—in differing forms, from Aristotelian theology to Kant's critique of reason to positivistic epistemology—of the relationship of grounding, of the appeal to the entity or to a supreme entity, first or last, that "grounds" it. As Heidegger has shown, especially in *Der Satz vom Grund*,[11] grounding conceived in this way belongs to metaphysics as forgetfulness of Being insofar as it grounds only to the extent that it is "returned" to the subject that believes he finds it outside of himself. The principle of sufficient reason is "*principium* reddendae *rationis*." This may lead one to think that nonmetaphysical thought, that is to say, a thought not forgetful of Being, would be able to open itself *truly* to alterity, and not to that false alterity of grounding "returned" to the subject, as though the limit of metaphysics were not above all that of reducing Being to the entity, but instead that of being unable to furnish a foundation solid enough (since what the subject believed to be external to itself is instead posited and established by the subject itself). To forgetfulness of Being defined as the reduction of everything to the power of the subject (finally, in the form of the technico-scientific will to power realized in the world of total organization), one would oppose a "recollective" thought able truly to meet with the "other," Being as irreducible to the subject and to its ability to order. If we were to think in this way, going beyond metaphysics—as certain pages of Heidegger seem to suggest, pages that have given rise to a sort of "rightist" version of Heideggerism—would end up as a radicalization of the "thought of the other" that finds its expression in Emmanuel Lévinas more than in Heidegger.

There are many reasons to consider Lévinas's thought a decisive contribution in the direction of the going beyond of metaphysics of which so much twentieth-century philosophy speaks. Above all, and in a way more radical than even Adorno, Lévinas carries out all the implications of the ethical necessity in whose name metaphysics seems to require a going beyond. Unlike Adorno, Lévinas develops this theme in the name of ethics while at the same time attempting radically to renew philosophical language. Adorno, on the other hand, while remaining within a dialectical perspective, was still completely tied to traditional conceptual language. Radicalization of the ethical necessity and the attempt to depart from the language of the metaphysical tradition are closely intertwined in Lévinas, even if, in the final analysis, the most significant result of his thought as far as going beyond metaphysics is concerned is probably not to be found so much in the "reduction" of metaphysics to ethics as in the problematization of the conceptual language of metaphysics in relation to another tradition to which he refers: the Jewish religious tradition. Lévinas himself only partially thematizes this second aspect of his thought, which opens the way to the problem of secularization as a decisive step in going beyond metaphysics. He seems to consider it marginal or "instrumental" (in the sense that it is in some way primarily of methodological interest) with respect to the reaffirmation of ethics as the only basis for a thought that would no longer be violent.

What Heidegger, Adorno, and much of twentieth-century critical thought call metaphysics and point to as a mode of thought to be gone beyond, Lévinas calls *ontology*. He reserves the term "metaphysics" for thought that, by opening itself to the beyond, to alterity, evades the logic of violence that has characterized traditional ontology (or metaphysics). He thus follows one of its historically consolidated meanings (*meta ta physika* = beyond physical things, the given visible world, etc.) Ontology is in fact the knowledge (*logos*) of Being as such, a knowledge that European

philosophy from Socrates to Heidegger has always considered the pre-
liminary condition of access to entities. Philosophy—or the prevailing
line in Western philosophy—has always held that the task of theory was
to appropriate to itself the notion of Being (ontology or metaphysics has
always been the science of Being as Being). When Heidegger declares
that ontic knowledge (knowledge of entities, of individual given things)
presupposes a knowledge or ontological fore-understanding, that is, of
Being as such; or when Husserl theorizes that the entity can give itself
only by rising against a background, against the horizon that exceeds it,
"as an individual arises from a concept,"[12] they merely repropose the old
ontological program of European philosophy. That program, like ontology, "has most often been a . . . reduction of the other to the same" (TI,
43). Individual entities, for that philosophy, can appear in experience only
as "individuals" against the background of a totality that renders them
comprehensible only to the extent that it brings them back to itself. Totality is Being that the knowing subject must always already know before
gaining access to the entity. The latter, however, insofar as it is brought
back to that already-known, to the totality of the universal concept, is not
grasped at all as that which it is, but is immediately reduced to the Same
for the knowing subject. The other, however, not so much as inanimate
object or nature as the other man, is not merely a specimen reducible to
the general notion, to totality. We must arrive at this conclusion if we pay
attention to the theme that opens one of Lévinas's major theoretical
works, *Totality and Infinity*: desire, a radical experience of alterity that, taken
as a point of departure, places Lévinas from the very beginning within an
original theoretical horizon, irreducible to the theoreticism prevalent in
the metaphysical tradition. There is a "metaphysical" component (in the
specifically Lévinasian sense) in desire that renders it irreducible to need,
or even to passion: desire cannot be satisfied like hunger or sexual im-
pulses. "The metaphysical desire tends toward *something else entirely*, toward
the *absolutely other*" (TI, 33). This absolutely other, however, is the other
man in his infinity, not only another "I" toward whom I am responsible in
a sort of equal relationship. In the other I seek not only a "thou" that
"completes" me, or with whom to share a common "nature," etc. The other
is desired insofar as he himself desires, that is to say, is open onto an
infinity. The other is a face, not only a visage,[13] because it looks at me
when I look at it and above all because it speaks to me.

> Speech cuts across vision. In knowledge or vision the object seen can
> indeed determine an act, but it is an act that in some way appro-
> priates the "seen" to itself, integrates it into a world by endowing it
> with a signification, and, in the last analysis, constitutes it. In dis-
> course the divergence that inevitably opens between the Other as my
> theme and the Other as my interlocutor, emancipated from the
> theme that seemed a moment to hold him, forthwith contests the
> meaning I ascribe to my interlocutor. . . . The idea of infinity, the infi-

nitely more contained in the less, is concretely produced in the form
of a relation with the face. (TI, 195–96)

Since the relation with the other is, according to Lévinas, the first
true experience of Being—what for Heidegger is instead fore-understanding—in Lévinas a constitutive relationship between Being and language—or better, between Being and discourse/dialogue—is established
as well. The other as face "produces" himself as infinity in the discourse
he addresses to me, and *this* strips it of all violence that it could exert on
the "Same," on the I who encounters it. "The relation with the other as
face . . . is desire, teaching received, and the pacific opposition of discourse" (TI, 197). The relation with the other as face is, according to
Lévinas, an ethical relation, but it "does not limit the freedom of the
same; calling it to responsibility, it founds it and justifies it" (TI, 197) because it is an asymmetrical relation: "The Other, in his signification prior
to my initiative, resembles God" (TI, 293). Even though one cannot say
simplistically that the infinity of the other consists in his being an image
of God—which would still be an ontological reduction of the other, re-
spected only "as" creature, image, etc.—the respect that he exacts is
exacted not only in the name of an equality or reciprocity of rights, but
also in the name of a transcendental dimension in which he is. Lévinas
describes this dimension of the infinity of the other by appealing to the
Cartesian formulation of ontological proof and the idea of infinity at its
base. In the idea of infinity that he illustrates in the third meditation, Descartes "discovers a relation with a total alterity irreducible to interiority,
which nevertheless does not do violence to interiority" (TI, 211); and in
the last paragraph of the same meditation, philosophy comes to conceive
of God as a person: "It is a question no longer of an 'infinite object' still
known and thematized, but of a majesty" (TI, 212).

In what is probably the best essay on Lévinas's thought, Jacques Derrida has set forth a series of arguments in order to maintain that, ultimately, if ontology is violence in the sense meant by Lévinas, then Lévinas himself with his philosophy of alterity cannot completely escape
this violence.[14] On the one hand, the encounter with the other requires
that he in some way reveal himself to us precisely "as" other, and hence
that he be thematized as an *ego* (and not, for example, as a thing), but this
involves at least a certain fore-understanding of an ontological sort.[15] On
the other hand, if, as Lévinas would have it, the face of the other presents
itself to me as an interlocutor in a discourse, the encounter will always
already be mediated by language, which is also ontological "violence":
"Predication is the first violence. Since the verb *to be* and the predicative
act are implied in every other verb, and in every common noun, nonvio-
lent language, in the last analysis, would be a language of pure invocation, pure adoration, proffering only proper nouns in order to call to the
other from afar."[16]

If Derrida is right, at least on these points, there are reasons to believe that the effort to go beyond metaphysics (as ontological violence) fails once again, even in Lévinas. One can probably add yet another argument to those of Derrida—an argument that merely makes explicit hints already found in Derrida's text. Without discourse and the relative violence that it implies, Derrida says, "there would be only pure violence or pure nonviolence."[17] It is easier to understand the second alternative and hence conclude not only that in spite of the effort to think Being in terms of pure alterity, Lévinas must accept a certain measure of "transcendental violence" outside of which even the other as other could not give himself, but also that perhaps precisely the other as other is, at bottom, a "figure" of violence, still a form of peremptory authority of the metaphysical *Grund*. At times this is explicit in Lévinas's own text: thus in one of the concluding pages of *Totality and Infinity*, the exteriority of Being is "entirely command and authority" (TI, 291), an infinity that turns to the Same as its highness and majesty. If, as we have learned from Heidegger, metaphysics is not only the violence of the reduction of everything under a universal, but also and inseparably the identification of this universal with an entity—grounding, *archē*, first principle, authority—then Lévinas appears as someone who, in order to escape from metaphysics in its first sense as ontology, simply rediscovers metaphysics in its second sense as theology. Even though this is an aporia that Lévinas himself seems unable to resolve, it is here that we find the most fecund aspect of his thought: that which sends us in the direction of the meditation on secularization and its meaning for going beyond metaphysics.

The link between metaphysics as the science of Being as Being and metaphysics as theology—the science of the entity that as a pure act realizes in itself Being in its fullness—is one of the most deeply rooted theses of Western thought. That Lévinas has recourse to the idea of infinity as it "functions" in Descartes shows that, at least in this respect, he does not refuse his own membership in the metaphysical tradition coming from the Greeks. But to that extent, it is also probable that his notion of alterity succeeds only with great difficulty in escaping from the implications of what Heidegger has called the onto-theology that runs through all of European thought. Even if, with his evocation of Descartes, Lévinas's "Other" seems to define itself within the onto-theological horizon and to reflect in itself, in the final analysis, even traits of violence (in the "command and authority" that characterize it), it is also true that in other pages of his work, and particularly in the preface to *Totality and Infinity* (the work in which the evocation of Descartes is most essential and constant), Lévinas seems to recognize as a "source" of his philosophy of exteriority, not phenomenological analysis carried out with the conceptual instruments of Western philosophy, but the eschatology of the biblical prophets. Eschatology does not mean primarily a certain anticipation of the last things that would present itself as the "completion" of philosophical evidence (TI, 22). It takes shape above all as an idea of judgment pronounced upon each being in every moment, outside the succession and unraveling of their destiny in history as totality. It is only insofar as they are "able" to be subjected to a judgment that puts them in relation to the beyond of history, that is, of totality, that men can have authority as "Others" and present themselves as faces open onto infinity. In the light of the "extraordinary phenomenon of prophetic eschatology," (TI, 22), Western philosophy appears dominated by the idea of Being as totality and by the violence that it implies: "We oppose to the objectivism of war [i.e., the link totality-violence of ontology] a subjectivity born from the eschatological vision" (TI, 25).

And yet Lévinas thinks he can carry out this undertaking with the instruments provided by Husserl's phenomenology (from which thought learns that one must first of all "let be" [TI, 29]) without, however, remaining within the "ontological" limits that mark it. He maintains that eschatology does not live on "subjective opinions and illusions" and that instead philosophical evidence itself "refers from itself to a situation that can no longer be stated in terms of totality" (TI, 24). Lévinas is well aware of the difficulties implicit in this situation: prophetic eschatology as true source of the thought of the other; the possibility that "Western" philosophy could manage on its own to dissolve the dominion of the notion of totality: "It is perhaps time to see in hypocrisy not only a base contingent defect of man, but the underlying rending of a world attached to both the philosophers and the prophets" (TI, 24). Is it possible that the other could truly escape from the mechanism of metaphysical violence insofar as he is thought not so much on the model of the Cartesian idea of infinity, but rather as the "Lord" who speaks in the Bible?

It seems that Lévinas's entire *opus*, even and especially the many edifying pages that appear in *Totality and Infinity*, is an effort to show that the "majesty" of infinity thought according to Biblical eschatology does not have the traits of violence that are instead proper to metaphysical Being. But isn't the Being of Western ontology perhaps violent because it "reduces" to itself? Is this reduction, though, characterized as violent only or principally for "theoretical" reasons? This would contradict the ethical inspiration of Lévinas's thought. The reduction to the same is called violence not because it does not allow the other to appear in that which it truly is, as though the essential thing were to know or make oneself known in one's own true nature, and the Good were not instead, as Lévinas often says, above truth and Being. The violence of ontology would then consist in the exercise of a power, of a command: exactly what Lévinas attributes to the alterity of the infinite-Lord.

The problem of the relationship between the *logos* of Western philosophy and biblical eschatology is intimately entwined with these problems of content. One can therefore say that the hidden center of Lévinas's thought is the problem of secularization. Lévinas seems to flee continually from this problem, at least on the surface. The "explosion," as he calls it in *Totality and Infinity*, of the category of totality occurs both because biblical eschatology calls us and because philosophy itself joins in uncovering its insufficiency. But doesn't this double inspiration perhaps constitute the hypocrisy of European thought that we should begin to conceive in terms that are not exclusively moralistic (cf. TI, 24)? Lévinas does not make clear whether and to what extent he considers Western philosophy to be *entirely* dominated by an orientation like that of Ulysses, who in the end returns to his native island, "une complaisance dans le Même, une méconnaissance de l'Autre,"[18] or whether biblical eschatology is, as it seems, truly a source completely other from that of thought dominated by the violence of the same. As we have seen, he does not openly theorize such a radical alternative. But at that point he should ask himself whether it is in general possible that philosophy on its own could manage to blow up the idea of totality that has always dominated it. One can see two possible schema here: either the thought of exteriority can be instituted through the theoretical "correction" of errors in Western thought, and then biblical eschatology would be only a *complementary* source of truth; or Greek philosophy is completely dominated by the idea of totality and to "correct itself" needs an "external" intervention like that of the word of the prophets. The first schema does not seem feasible, especially because to give the thought of totality the form of an "error" able to "correct itself" would mean to continue to adopt a "representative" image of theory: we believed erroneously that things stood in a certain way, but now we discover, etc. But to represent, and hence also to "correct," a certain erroneous representation with a more adequate one is precisely the procedure of the reduction of the other to the same. It is difficult to think that this could be the way to get out of the violent ontology of totality. What remains, then, is the second schema: external intervention. But what Lévinas verifies and Derrida makes quite clear is that thought cannot succeed in getting completely out of the Western *logos*, which appears to Lévinas to be "the means of every agreement, in which all truth is reflected" and to be "Greek civilization and what it has produced: *logos*, the coherent discourse of reason."[19] What could the correct way to stay within the "laceration" between Greek *logos* and eschatologism be? That the two cannot be considered sources of truth completely separated appears in the very "necessity" in which Lévinas finds himself enveloped of having recourse to an element of traditional metaphysics like the idea of Cartesian infinity in order to "speak" of the exteriority of the other. These two languages are always already given to us in a mutual interwining, like a sending off and a destiny—in the double sense that the term *Ge-Schick* has in Heidegger.

Such a problematic seems to point to the following conclusion: only by thematizing more explicitly—and not as a marginal problem of language and method—the question of the relation between the two traditions to which he refers (Greek and biblical) can Lévinas get to the heart of the problem of ontological violence still implicated in his representation of the other. It is probably not without importance, here, that Lévinas refers to the Bible as Old Testament, to Judaism, and not to Christianity. This accentuates the tendency in him, visible in Christian thinkers as well, to solve the problem of secularization with a pure and simple "return" to the origin, to a moment preceding the "dissolution" that secularization presumably represented. What Heidegger calls the metaphysical forgetfulness of Being in favor of the entity, and Lévinas describes as the predominance of ontological violence, would be dissolved by a "recollection" that would constitute a step backward toward an authentic "representation" of Being as other, obscured by the contaminating itinerary to which Western history has submitted it. But wouldn't this return not be first of all purely and simply a rediscovery of the metaphysical *Grund* in a still-barbaric form, still marked by the peremptoriness of a relationship of command, whereas the metaphysics of "grounding" and finally the *principium reddendae rationis* represent instead a first secularization but also a civilizing effect, a reduction of violence?

It is certainly undeniable that the foundation of the contemporary problematic of going beyond metaphysics is ethical and, even more, religious. It is religious in one of the senses that Lévinas has brought to light: in seeking to go beyond metaphysics one encounters the "Lord" of the Bible whose alterity, however, consists in not being reducible to *Grund*. Not in the sense, as Lévinas seems to think, that *Grund* does not "ground" enough, that it is not other enough to be able to provide the base, authority, and command that desire would seek (thereby contradicting, however, his very own "infinity"), but rather in the sense that it puts us in relation to an "origin" that escapes every grounding mechanism. For this reason Lévinas calls that alterity "immemorial."[20] If we must take this immemoriality seriously, however, the Lord of the Bible cannot be thought of as the "creator," at least insofar as creation tends to take the form of an absolute beginning, but instead as perhaps merely the author of the message, or better, as the transmission of the messge itself. It is not true, as Lévinas believes, that biblical eschatology speaks above all of judgment that touches everyone at every moment apart from any historical "belonging," or better, that this judgment is exercised through belonging to the events of which eschatology speaks: creation, sin, redemption, expectation of the end of the world. Biblical eschatology is also a philosophy of history in which the secularization that Lévinas does not thematize is "foreseen" or begun. Moments of history are not all equal before God—a God for whom this would be true would be precisely nothing other than the God of philosophers, the first principle of metaphysical ontology. Derrida is right to chide Lévinas for a tendency to think alterity as a leap outside history,[21] even if Derrida then believes he can resolve the problem of historicity by making a philosophy of the structural finitude of man out of the inevitability of ontological violence (in the senses noted above). This philosophy as such would open the field of historicity, of choices, of human affairs, in which, however, it would still hold that all moments exist in an "immediate" relation to, and are equal with respect to the other, Being, God: hence a very sui generis historicity also unable to truly think secularization.

If the Lord of the Bible must not be a metaphysical "principle" grasped only in a more initial and barbaric configuration, the eschatology of the prophets can be considered neither a source parallel to that of Greek philosophy nor a truer preceding moment to which one should return by skipping over the equivocations and distortions introduced by the Greek *logos*. The "uses" to which the biblical message has been put, both Old and New Testaments—the Christianization of the West, "Weberian" rationalization as interpretation and "application" of the Bible—are all *part* of the biblical message, are completely internal to it. The "Lord of the Bible" is such in the double sense of the genitive, to use once again one of Heidegger's "grammatical" ploys: he is not only the author to whose will, personality, intention, or, in short, "presence" one can return through the text; he is also, inseparably, an effect of the text, the "continuity" that speaks to us in the interpretations, translations, and transmissions that constitute the history of Greco-Judaic-Christian civilization. To thematize this relation between the text and its author: precisely this and only this is, in the end, going beyond metaphysics.

Like Adorno with his notion of micrology, Lévinas says more than his theory actually manages to contain insofar as he does not carry out the thematics of secularization to its end. Thus in the beginning of *Autrement qu'être*, he writes: "Si la transcendance a un sens, elle ne peut signifier que le fait, pour l'*événement d'être*—pour l'*esse*—pour l'*essence* de passer à l'autre de l'être. . . . Passer à l'*autre* de l'être, autrement qu'être. Non pas *être autrement*, mais *autrement qu'être*. Ni non plus ne-pas-être. Passer n'équivaut pas ici à mourir."[22] To discover that the "Lord of the Bible" is not other inasmuch as he is authority, principle, grounding, but inasmuch as he is the author of a message that reaches us essentially marked by the itinerary of its transmission—and hence also by the fact that it has been understood as *Grund*, as sufficient reason, and finally as will to power in Nietzsche—and that as author he is not only origin but also always an effect of the text means to open oneself to a notion of Being that takes leave from metaphysics because it sees it in its constitutive connection with passage. One does not get back to the *archē* or to the creator or to the author. And this is the same as realizing that Being *is* not but *takes place*. The experience of secularization, the encounter (whether hypocrisy or laceration) between the Bible and Greek *logos* that constitutes the West, is not, on the level of the history of ideas, an aspect parallel to the coming to light of Being as an event. It is the same thing. The true meaning that the appeal to the biblical tradition has for going beyond metaphysics is that in such an appeal Being gives itself as an event (message, word transmitted) to which the history of metaphysics belongs as a constitutive moment. To go beyond metaphysics or, as Heidegger prefers, *verwinden* (to take up again, to distort, to submit to and recover from) metaphysics, is to travel again thematically the itinerary of its transmission or "to correspond to *Ge-Schick*," to its sending off, recognizing it in its nature as secularizing process.[23] If Lévinas's path remains that of the metaphysically oriented return to the other as initial moment, authority, and command, Heidegger's path consists in following the movement of secularizing "dissolution" in which Being, even through the contamination of the "Lord of the Bible" by Western metaphysics, liberates itself from its violent connotations. From "principle" it becomes word, discourse, interpretation. In the Heideggerian "reconstruction" of the history of metaphysics as the destiny of Being (which, for now, is accomplished in the Western sunset toward which metaphysical Being moves), Judaism and Christianity are strangely almost absent. Although he described metaphysics as onto-theo-logy, one could say that the recollection that he practiced stopped at the ontological aspect. The reception of Heidegger's teaching has something essential to learn from Lévinas's thought on this point. A more authentic understanding of the "event" of Being, outside of the recurrent temptations to get out of metaphysics by taking up theology in place of ontology, can probably arise only from an explicit elaboration of the Heideggerian recollection of metaphysics that would include its theological aspect. In its "theoretical" and, inseparably, its "epochal" aspects (*Ge-Stell*), the *Verwindung* of metaphysics is nothing other than secularization.

TRANSLATED BY BARBARA SPACKMAN

GIANNI VATTIMO
Toward an Ontology of Decline

According to a well-known thesis of Heidegger, the name "Occident," *Abendland*, not only designates our civilization's place in a geographical sense but names it ontologically as well insofar as *Abendland* is the land of the setting sun, of the sunset of Being. One can speak of an ontology of decline and see its preparation and first elements in Heidegger's texts only if one interprets Heidegger's thesis on the Occident by transforming its formulation: not "the Occident is the land of the sunset (of Being)," but "the Occident is the land of the sunset (and hence, of Being)." Moreover, another decisive Heideggerian formula that serves as the title of one of the sections of *Nietzsche*, "metaphysics as the history of Being," can be read in exactly the same way if accented correctly, that is to say, in the only way that conforms to the whole of Heidegger's thought: not "metaphysics is a history of Being," but "metaphysics is *the* history of Being."[1] Apart from metaphysics, there is no other history of Being. Thus the Occident is not a land in which Being fades, while elsewhere it shines (shone, will shine) high in the noonday sky; the Occident is the only land of Being precisely insofar as it is also, inseparably, the land of the sunset of Being.

In its intentional ambiguity, this reformulation of Heidegger's pronouncement on the Occident means immediately to take its distance from the most widespread interpretations of the meaning to be given Heidegger's philosophy. Generally speaking, these interpretations can be characterized as alternately emphasizing one or the other term, sunset or Being, to the detriment of the connection between them that seems to me indissoluble. Those interpretations that persist in reading Heidegger as a thinker who, in some way however problematic or preparatory, foretells a return of Being or to Being, according to a line of thought that may be called religious (or more precisely, theo-logical, in the sense of the Onto-theo-logy of *Identität und Differenz*), put exclusive emphasis on the term "Being."[2] Those interpretations that find in Heidegger's thought an invitation to take note that metaphysics has come to an end, and with it any possible history of Being, emphasize instead the term "sunset": of Being "nothing remains" at all, and this excludes any mythic expectation of its possible returning to us. The very liveliness with which these two readings, with all their internal differentiations, continually oppose and contend with each other may be taken as a sign that the two elements that they isolate, and that the proposed formula attempts to express in their connection, are indeed present and problematically connected in Heidegger's text. Even on a first and superficial reading, such a formula may begin, without forcing the matter, by explaining that which always seems an ambiguity in Heidegger's attitude toward the history of metaphysics. This ambiguity may be eliminated only by interpreting that history as a dialectical preparation of its own supercession in the direction of a recollective thought like that which Heidegger himself seeks to actualize. But *Metaphysik*, like the *Geschichte des Seins*, is precisely not a dialectical unfolding. The attention and respect—or better and definitely, the *pietas*—that Heidegger displays in regard to the history of that thought in which ever more clearly nothing remains of Being is not justified dialectically by an identification of the real (the event) with the rational. Rather, this *pietas* can be better explained by the awareness that metaphysics is the destiny of Being also and above all in the sense that decline "befits" Being.

Along with this, however, it has also been said that in Heidegger's texts one finds the premises and elements for a possible "positive" conception of Being, and not only the description of a condition of absence that would be defined only in relation—a relation of nostalgia, expectation, or even of liquidation (as in the repudiation of metaphysics as myth and ideology)—to the presence of Being understood as characterized by all the *strong* attributes that Western tradition has always conferred upon it. These attributes are strong not merely in a metaphorical sense. The relation between *energeia*, the actuality that characterizes Aristotelian Being, and *enargeia*, the evidence, luminosity, vividness of that which appears and imposes itself as true, exceeds that of verbal resonance, as does the relation between actuality and energy, and energy and true force. When Nietzsche speaks of metaphysics as an attempt to master the real by force, he does not describe a marginal characteristic of metaphysics but indicates its essence as it is delineated right from the very first pages of Aristotle's *Metaphysics*, where knowledge is defined in relation to the possession of first principles.[3]

I do not think that interpreters and followers of Heidegger have yet developed even the first elements of an ontology of decline, except in certain respects for Gadamer's hermeneutics, and the thesis "Being that can be understood is language."[4] There, however, the relation of Being to language is always studied predominantly from the point of view of language, and not from the point of view of the consequences that it might have for ontology. In Gadamer, for example, the Heideggerian notion of metaphysics receives no significant elaboration. The absence of a theoretical elaboration of the ontology of decline by members of the Heideggerian school probably depends on their continuing, in spite of every *Warnung* to the contrary, to think of Heidegger's meditation on Being in

terms of grounding. Heidegger, instead, made it necessary to "leave behind Being as grounding" if one wants to move on to recollective
thought.[5] If I am not mistaken, Heidegger spoke of *Fundamentalontologie*
only in *Sein und Zeit*, and though his texts speak often of *Begründung*, it is
always in reference to metaphysics, precisely the thought that moves only
against the horizon of the assignation of *Grund*. In *Sein und Zeit* a certain
intention of grounding does reveal itself, at least in a broad sense, for it
deals with an interrogation of the meaning of Being, that is, of the horizon
only within which every entity gives itself as something. But from the very
start, with the importance assumed by the reference to the passage from
the *Sophist* in the epigraph of the work, the inquiry is immediately oriented
toward a historical condition. Not even for a moment does it turn to pure
conditions of possibility, whether of the phenomenon or of knowledge, in
a Kantian sense. If we may play upon words, we are faced with a situation
in which the condition of possibility in the Kantian sense reveals itself to
be indissolubly connected with a condition understood as the state of
things, and this connection is the authentic topic of discourse.

In *Sein und Zeit*, we neither seek nor find what the transcendental conditions of the possibility of the experience of the entity might be, but we
ascertain in a meditative mode the conditions in which, in fact, our experience of the entity alone gives itself. Of course, this does not imply a
total abandoning of the transcendental plane, of the interest in identifying conditions of possibility in the Kantian sense. But the inquiry must
note from the very beginning that it can carry on only in an inextricable
connection with the identification of conditions in the factical sense of
the word. It is necessary to draw attention to this point, especially in relation to recent revivals in hermeneutical circles (such as those by Karl Apel
and Jürgen Habermas) of orientations for the most part Kantian. One of
the elements that already in *Sein und Zeit* constitutes a basis for the ontology of decline is precisely the specific physiognomy that "grounding"
takes on there. Precisely on account of the radical way in which the question of Being is posed in that work—with the immediate passage to the
existential analytic—it is clear that any possible reply to the question
can, in principle, no longer take shape as grounding not only in the sense
of the assignation of *Grund*, of sufficient reason or principle, but also in
the sense that thought cannot in any way expect to reach a position from
which to have at its disposal the entity that is supposed to be grounded.
Already in *Sein und Zeit* Being is "left behind as grounding." In the place of
Being able to function as *Grund* one glimpses, in the centrality that the
existential analytic and the elucidation of the link with time assume, a
"Being" that is constitutively no longer capable of grounding: a weak and
depotentiated Being. The "meaning" of Being that *Sein und Zeit* seeks and,
to some extent, attains must be understood above all as a "direction" in
which being-there and the entity find themselves headed: a movement
that leads them not to a stable base but to a further, permanent dislocation in which they find themselves dispossessed and deprived of any

center. The situation that Nietzsche describes (in the note that opens the old edition of *Wille zur Macht*) as characteristic of nihilism, in which from Copernicus on "man rolls away from the center toward the X," is also the situation of Heideggerian *Dasein*. *Dasein*, like post-Copernican man, is not the grounding center, not does it inhabit, possess, or coincide with that center. The search for the meaning of Being, in its radical development in *Sein und Zeit*, makes it progressively clear that this meaning is given man only as dispossession and un-grounding. Hence even against the letter of Heidegger's texts, one must say that the search begun in *Sein und Zeit* does not send us in the direction of going beyond nihilism, but rather of experiencing nihilism as the only possible path of ontology.

This thesis runs against the letter of Heidegger's texts because nihilism there means the flattening out of Being onto entities, the forgetfulness of Being that characterizes Western metaphysics and that, in the end, reduces Being to "value" (in Nietzsche), to validity posed and recognized for and by the subject. So it happens that, of Being as such, nothing remains. This is not the place to discuss whether and to what extent nihilism understood in this way faithfully and completely characterizes Nietzsche's position. It is clear, however, that Heidegger's use of the notion of nihilism to indicate the culmination of forgetfulness of Being in the final moment of metaphysics is responsible for the fact that, from his thought as alternative or as attempt to go beyond and as opposed to what occurs in the case of nihilism, one expects Being to recuperate its function and its grounding force. Instead, precisely this grounding force and function also still belong to the horizon of nihilism: Being as *Grund* is only a preceding moment of the linear development that leads to Being as value. Of course this is well known to readers of Heidegger, but it is a matter of meditating upon it again and again in order to draw its considerable consequences. The peculiar link between grounding and un-grounding in *Sein und Zeit* means that, in the final analysis, the search for the meaning of Being cannot give rise to the attainment of a "strong" position, but only to the taking on of nihilism as a movement by which man, *Dasein*, rolls away from the center toward the X.

The link grounding/un-grounding runs through all of *Sein und Zeit* and emerges especially in such moments as the inclusion of *Befindlichkeit*, the emotive situation, among the existentials, that is, among the constitutive modes of the opening of *Dasein* which, in Heidegger, "substitute" for the Kantian transcendental; or in moments like the description of the hermeneutic circle in the light of which truth appears tied to interpretation as elaboration of the fore-understanding into which being-there is always already thrown by the very fact that it exists; and, above all, in the constitutive function that being-unto-death exercises in the face of the historicity of being-there. The function and import of being-unto-death is precisely one of the points most resistant to interpretation and to theoretical recuperation and elaboration in the entire *Sein und Zeit*. Authoritative interpreters such as Hans Georg Gadamer, for example, put into doubt its very systematic connection to the whole of Heidegger's thought. Even structurally, the discourse on being-unto-death is exemplary of the way in which *Sein und Zeit* sets off in search of a still metaphysical ground and arrives at nihilistic results (at least in the sense to which I have alluded above).

Heidegger in fact arrives at being-unto-death by posing a problem that at first seems perfectly "metaphysical" in form and content: has the existential analytic of the first section put *Dasein* at our disposal in the totality of its structures?[6] But, Heidegger asks himself, what does it mean for being-there to be a totality? Coherently followed through, this problem leads him to see that being-there constitutes itself in a totality and hence "grounds" itself to the extent that it anticipates its own death (since the assignation of *Grund*, of which grounding consists, has always meant the closure of the series of connections, and the constitution of precisely such a totality against the regression *ad infinitum*). Freely translating Heidegger's language, we can say: being-there is truly *there*, that is, it distinguishes itself from beings-in-the-world insofar as it constitutes itself as historical totality that goes along continuously, historically, among the various possibilities that, by coming into being or disappearing along the way, make up its existence. Even inauthentic Being, as a simple defective mode of historical existence as continuity, harks back to being-unto-death: its constitutive category is always death but experienced in the form of *man*, of the quotidian "one dies." The constitution of being-there in a historical continuum has to do radically with death insofar as death, as the permanent possibility of the impossibility of all other possibilities, and hence as authentic possibility insofar as it is authentically a possibility, lets be all the other possibilities on this side of itself and maintains them in their specific mobility, prevents them from rigidifying into exclusive possibilities-realities and allows them instead to constitute themselves in a texture-text.

All of this means, however, that being-there exists and thus acts as the place of illumination of the truth of Being (that is, of the coming to Being of entities) only insofar as it is constituted as the possibility of no longer being-there. Heidegger insists that one must not read this relation to death in a purely ontic sense, and therefore not even in a biological sense. Nevertheless, like all moments in which philosophy encounters analogous points of passage (above all, that between nature and culture), this Heideggerian distinction is dense with ambiguity. If it is in fact true that being-there is historical—that is, it has an existence as continuous *discursus* endowed with possible meanings—only insofar as it can die and explicitly anticipates its own death, it is also true that it is historical in the sense that it has at its disposal determined and qualified possibilities and has relationships to past and future generations precisely because it is born and dies in the literal, biological sense of the words. The historicity of being-there is not only the constitution of existence as texture-text; it is also the belonging to an epoch, *Geworfenheit* that intimately characterizes the project within which being-there and entities hark back one to the other, come to Being in ways shaped differently each time. This double meaning of historicity, in its relation to being-unto-death, is one of the points at which, most explicitly though problematically, the link grounding/ungrounding comes to light; this link is one of the meanings, and perhaps *the* meaning of *Sein und Zeit*.

Whether and to what extent the elucidation of this link also involves, as it seems to me, renewed attention not only to the ontological but also to the ontic, biological meaning of death, is a question that should be addressed elsewhere. What interests us here is to show that the Being toward which Heidegger speaks can no longer be thought with the characteristics of metaphysical Being, not even when it is qualified as *hidden* or *absent*. It is therefore false and misleading to think that Heideggerian ontology is a theory of Being as force and luminosity obscured by some catastrophic event or even by a limitation internal to Being itself, by its epochality, and that it wants to act as preparation for a "return" of Being, still understood as luminosity and grounding force. Only if one thinks in this way will one be scandalized by the thesis according to which *the outcome of Heidegger's meditation, starting with* Sein und Zeit, *is the taking on of nihilism*, which, in the "un-grounding" sense in which Nietzsche experiences it in the note cited from *Wille zur Macht*, is a current present but not dominant in the metaphysical tradition, which instead has always moved according to the logic of *Grund*, of substance and value. To recognize completely the implications of this Heideggerian nihilism (and we are only at the beginning) means, for example, to close the door upon the interpretations of his thought in terms, explicit or implicit, of "negative theology," both those interpretations that take him as the theoretician of *dürftige Zeit* that looks back with regret upon and awaits the "strong" giving itself of Being (as presence of transcendent Being, for example, or as decisive historical event that would open a new history to a no-longer alienated man) as well as those that read his announcement of the end of metaphysics as the freeing of a space for an experience that would be organized independently of Being (once again, still characterized as a gravity of a metaphysical sort). In the interpretation proposed here, the outcome of Heidegger's thought is not the assertion that the grounding guaranteed by metaphysical Being does not give itself (anymore, or yet) and that thought must consequently look back upon it with regret or prepare its coming. Nor does it consist in taking note that such grounding is finally rendered vain and that consequently we can and must go on to construct a "non-ontological" humanity turned exclusively toward entities and engaged in the techniques of organizing and planning their different domains. The latter position lacks (like the former, in fact) a critique of the "strong" conception of Being and rediscovers that conception without recognizing it insofar as it ends up attributing to entities and their domains the same peremptory authoritativeness that the thought of the past attributed to metaphysical Being.

We must therefore continually rethink—as in a sort of therapeutic exercise—the link grounding/un-grounding that announces itself in *Sein und Zeit* and runs through the entire later development of Heidegger's works. Not only does it manifest itself in the ambiguity of being-unto-death, but it also alludes to a "nontranscendental," and therefore also not "strong" in the metaphysical sense, relation between "right" and "fact" that opens the way to a completely new conception of the very notion of grounding. *Sein und Zeit* began the search for the meaning of Being as though it were a matter of identifying a transcendental "condition of possibility" of our experience. But immediately the condition of possibility is revealed as also the historical-finite "condition" of *Dasein*, which is rather project (hence a sort of transcendental screen) but *thrown* project, characterized each time by a different fore-understanding rooted equiprimordially in its emotive situation, in *Befindlichkeit*. The grounding that in such a way is not "reached" but rather "delineates itself" (since it is never something like a fixed point at which one arrives in order to stop there) may be defined with an oxymoron: *hermeneutic grounding*. Since it functions by grounding only (more) in this sense, Being takes on a connotation wholly foreign to the metaphysical tradition, and it is precisely this that the formula "ontology of decline" means to express.

The idea of a hermeneutic grounding appears in Nietzsche before Heidegger, and not by chance if both thinkers are moving within the horizon of nihilism. Read, for example, aphorism 82 of *The Wanderer and His Shadow* entitled "An Affectation in Parting":

> He who wishes to sever his connection with a party or a creed thinks it necessary for him to refute it. This is a most arrogant notion. The only thing necessary is that he should clearly see what tentacles hitherto held him to this party or creed and no longer hold him, what views impelled him to it and now impel him in some other directions. We have not joined the party or creed on strict grounds of knowledge. We should not affect this attitude on parting from it either.[7]

Is it only a question here of a call to the "human, all-too-human" roots of all that we consider validity and value? That too, probably. But one grasps the sense of this aphorism completely only when it is linked to the announcement that "God is dead," an announcement that is at once the "truth" that grounds the thought of un-grounding (there is no longer a strong metaphysical structure of Being) and the recognition that this "truth" can, in a particular sense, only be a statement of fact.

To understand this hermeneutic grounding as a pure and simple profession of historicist faith would mean to move again within the horizon of the metaphysical meaning of Being which, by its presence elsewhere or its pure and simple absence, continues to devalue all that is not "grounded" in a strong sense, making it fall into the realm of appearance, of the relative, of de-value. The historical-finite thrownness of *Dasein* never, however, allows an overturning of the existential analytic onto the level of the identification of historical-banal characteristics of epochs and societies, since to radicalize the historicity of the thrown project leads to calling into question the claims of a historicist grounding, and to repropose the problem of the very possibility of historical epochs and humanity on the level of the *Geschick* of Being. A radicalization of the historicity of the thrown project and placement of the problem onto the level of the *Geschick* of Being is what happens in the turn, in the *Kehre* of Heideggerian thought beginning in the 1930s.

But the *Kehre* does not allow itself to be reduced to a more or less veiled recuperation of historicism *only* if one clearly identifies in it the procedure of hermeneutic grounding which brings as its corollary the explicit enunciation of an ontology of decline. The meaning of the *Kehre* is the coming to light of the fact that thinking means grounding, but that grounding can have only a hermeneutic sense. After the *Kehre*, Heidegger incessantly goes over the paths of the history of metaphysics again and again, adopting that "arbitrary" instrument par excellence (at least from the point of view of the exigencies of the grounding rigor of metaphysics): etymology. What we know about hermeneutic grounding is, finally, all here. Entities give themselves to being-there within the horizon of a project which is not the transcendental constitution of Kantian reason but historical-finite thrownness that unfolds between birth and death, within the limits of an epoch, a language, a society.

The "he who throws" of the thrown project, however, is neither "life" understood biologically, nor society or language or culture; it is, says Heidegger, Being itself. Being has its paradoxical positivity precisely in not being any of these supposed horizons of grounding and in putting them instead in a condition of indefinite oscillation. As thrown project, *Dasein* rolls away from the center toward the X; the horizons within which entities (including *Dasein*) appear to it are horizons that have roots in the past and are open toward the future, that is to say, they are historical-finite horizons. To identify them does not mean to have them at one's disposal, but to be always sent to further links, as in the etymological reconstruction of the words of which our language is made. This hermeneutic retracing *ad infinitum* is the meaning of Being that *Sein und Zeit* sought. But this meaning of Being is precisely something totally different from the notion of Being that metaphysics has handed down to us. Before Heidegger and Nietzsche, the history of thought offers only one other decisive example of a theorization of hermeneutic grounding: the Kantian deduction of judgments of taste in the *Critique of Judgment*. There as well, grounding (in the specific case of the peculiar universality of judgments of the beautiful) comes down to recalling the subject's belonging to humanity, a belonging that is problematic and always in the process of becoming, just as "humanity," equated with the *sensus communis* to which judgment of taste refers, is also problematic and always in the process of becoming.

The most significant document that the mature Heidegger's work supplies for beginning to think in terms of hermeneutic grounding in a more articulated way seems to me to be his meditation on the essence of technology and the notion of *Ge-Stell*. A thesis like that which Heidegger enunciates in *Identität und Differenz*, according to which "in the framing [*Ge-Stell*], we glimpse a first, pressing flash of the appropriating event [*Er-eignis*]," may, without exaggeration, be compared to the Nietzschean announcement of the death of God, for it is close to it in many senses, both in content and in the way in which it establishes its own validity.[8] As in Nietzsche's "God is dead," we are here faced with the announcement of a grounding/un-grounding event: grounding, insofar as it defines and determines (in the sense in which *be-stimmt* indicates also "to in-tone") the condition (the possibility, the fact) of the coming of entities to Being; un-grounding, because this condition is defined and determined precisely as lacking all grounding in the metaphorical sense of the term.

Ge-Stell, as is well known, is the term with which Heidegger indicates modern technology as a whole, its *Wesen* in the contemporary world as an element that determines, *be-stimmt*, the horizon of *Dasein*. In Italian, the term *Ge-Stell* is translated as "im-position," written with a hyphen, in order to make visible the original sense of *Stellen*, to place, as well as the sense "impose" and the coerciveness that Heidegger attributes to it. What is lost in the translation is the meaning of the collective prefix, *ge-*, that indicates the totality of placing (but the coerciveness to which "im-position" alludes is perhaps also the most evident and fundamental trace of the meaning of the "totality" of technological placing).

As the totality of the world of technology, *Ge-Stell* defines the condition (the situation) of our specific historical-finite thrownness. It is also the condition of possibility of the coming to Being of entities in this determinate epoch. This condition of possibility is not open only in a "descendant" sense, like any other condition of possibility. It does not only make each entity appear as that which it is (*als etwas*), but is also the flash of *Er-eignis*. *Er-eignis* is another key term in the late Heidegger, which literally means "event" but is used by Heidegger with an explicit reference to *eigen*, one's own, to which it is linked. *Er-eignis* is thus the event in which each entity is "propriated," and hence appears as what it is insofar as it is also, inseparably, involved in a movement of "transpropriation." The movement of transpropriation concerns man and Being before things. In *Er-eignis*, in which entities come to Being, it happens that man is *ver-eignet* (appropriated) to Being, and Being is *zugeeignet* (consigned) to man.[9]

What does it mean, then, that in *Ge-Stell*, that is, in the im-position of the world of technology, this game of appropriation-transpropriation, of which the event of Being consists, shines forth? The fact is that *Ge-Stell* as totality of placing is not characterized only by planning and the tendentious reduction of everything to *Grund*, to a grounding foundation, and hence to the exclusion of all historical newness. Like the totality of placing, it is also essentially *Heraus-forderung*, pro-vocation; in the world of technology, nature is continually provoked, called upon to serve ever-new purposes, and man himself is called over and over again to involve himself in new activities. If, then, on the one hand technology seems to exclude history insofar as it is tendentiously planned, on the other hand, this "immobility" of Ge-Stell has a vertiginous character in which a continual, reciprocal provocation is in force between man and things, a provocation that may be designated by another Heideggerian term, round dance, the Reigen to which the final page of the essay "The Thing" links the Gering of the world (with the meaning of "lowest," but also "ring" and "to struggle" Ge-ring) as the Geviert, the fourfold.[10] Ge-Stell places being-there in a situation in which "our whole human existence everywhere sees itself challenged—now playfully and now urgently, now breathlessly and now ponderously—to devote itself to the planning and calculating of everything."[11] All of this urgency of the technological provocation in which our historical existence is—wesentlich—thrown may also be called shock (there are possible references to Georg Simmel, as well as to the shock of art in Walter Benjamin).

In these same pages of Identität und Differenz to which I refer, Er-eignis is defined as "that realm, oscillating within itself, through which man and Being reach each other in their essence, achieve their active essence by losing those determinations with which metaphysics has endowed them."[12] The determinations that man and Being have had in metaphysics are, for example, those of subject and object or, as Heidegger emphasizes a few pages later, those which have determined the twentieth-century distinction between the science of nature and the science of the spirit, between "physics" and "history," that is, the division between a realm of spiritual freedom and a realm of mechanical necessity.[13] In the round dance of Ge-Stell, precisely these opposed determinations are lost: things lose their rigidity insofar as they are totally absorbed in the possibility of total planning and are pro-voked to ever-new uses (without any more reference to a supposed natural "use-value"), and man also becomes not only subject but also always a possible object of universal manipulation.

All of this represents not merely the demonic import of technology. It is, instead, precisely in its ambiguity, the flash of Er-eignis, of the event of Being as the opening of a realm of oscillation in which the giving of itself of "something as something," the "self-propriating" of entities each in its own definiteness, happens only at the price of a permanent trans-propriation. Universal manipulability—of things and being-there—liquidates the characteristics that metaphysics had attributed to Being and to man, above all the stability (immutability, eternity) of Being to which is opposed a realm of freedom that is problematic and in the process of becoming. To think the essence of technology, as Heidegger says, and not only technology as such thus probably means to experience the provocation of universal manipulability as a call to the event-character of Being.

In the first of the two essays that make up Identity and Difference, "The Principle of Identity," there is a dense network of connections among the description of *Ge-Stell* as the place of the urgency of provocation, the description of *Er-eignis* as realm of oscillation, and a notion that (as the second essay, "The Onto-Theo-Logical Constitution of Metaphysics," shows) is central to the last phase of Heidegger's thought: the notion of *Sprung*, springing (to which is related also the notion of *Schritt-zurück*, the step back). Thought that "leaves behind Being as grounding" (following the formulation of *Zur Sache des Denkens*) in the sense of hermeneutic grounding is one that abandons the metaphysical realm of representation in which reality is disclosed in an order of dialectical mediations and concatenations and, precisely insofar as it evades this grounding chain, springs away from Being understood as *Grund*.[14] This springing must lead us, Heidegger says, to where we already are, in the constellation of man and Being represented in *Ge-Stell*. The leap does not find, upon arrival, a base upon which to land but only *Ge-Stell* as the place in which the eventuality of Being shines forth and makes itself accessible to us as a realm of oscillation. Being is not one of the poles of this oscillation, which instead moves between being-there and entities; it is the realm, or the oscillation itself. *Ge-Stell*, which can represent the greatest danger for thought because it throroughly develops the implications of the metaphysical rigidification of the subject-object relation in technology as total organization, is also the place of the flash of *Er-eignis* because the universal manipulability, the provocation and shock that characterize it constitute the possibility of experiencing Being outside of metaphysical categories, above all, that of stability.

Why can the experience of *Ge-Stell*, thus summarily described, take the form of an example of "hermeneutic grounding"? Here we rediscover the two elements that constitute Nietzsche's "God is dead," because (1) *Ge-Stell* is not a concept; it is a constellation of belonging, an event that *be-stimmt* any possible experience of the world for us, and functions as grounding insofar as, like "God is dead," one receives its announcement, and (2) belonging to *Ge-Stell* functions as grounding only inasmuch as it gives access not to a *Grund* "absolutum et inconcussum," but rather to a realm of oscillation in which every propriation, every giving of something as something, is suspended from a movement of transpropriation. The hermeneutic character of grounding that is thus put into effect seems to be linked above all to the first of these two aspects: inasmuch, that is, as one takes note that the conditions of possibility of our experience of the world are historical-finite, a historically situated fore-understanding. But isolated from the second aspect, this "grounding" would be only an overturning of the Kantian transcendental into historicism. The genuinely hermeneutic character of grounding is assured instead by the second of the two aspects, which, if you like, is the nth metamorphosis of the hermeneutic circle of *Sein und Zeit*.

Access to *Er-eignis* as realm of oscillation is made possible not by technology but by listening to its *Wesen*, which must be understood not as essence but as "to be in force," the way of giving itself of technology.

To think not technology but its *Wesen* requires that step back that Heidegger speaks of in the second essay of *Identity and Difference* (and that corresponds to the springing of the first essay) and that places us before the history of metaphysics in its totality. One of the difficulties that we encounter when explicating the meaning of technology and of *Ge-Stell* in Heidegger (ending of metaphysics but also the flash of *Er-eignis*) is that his text does not explicate further in what sense thinking the essence of technology, and hence experiencing *Ge-Stell* as the flash of *Er-eignis*, implies also a placing of oneself before the history of metaphysics in its totality but *not* from the point of view of a dialectical representation of that history.[15] It is legitimate to try to fill this blank by referring to another text in which Heidegger also speaks of springing: the pages of *Satz vom Grund* where we read that the principle of reason calls us to spring away from *Grund* into the *Abgrund*, the abyss, that is at the bottom of our mortal condition. We accomplish this springing to the extent that "we entrust ourselves, recollecting, to the liberating link that places us within the tradition of thought."[16] Access to the realm of oscillation thereby acquires a further and more explicit hermeneutic character; to respond to the call of *Ge-Stell* also involved a springing that puts us in a liberating relation to *Überlieferung*, that game of the transmission of messages and words that makes up the only element of possible "unity" of the history of being (which is completely resolved in this transmission of messages). Nietzsche had polemically described nineteenth-century man as a tourist who roams in the garden of history as though in a warehouse of theatrical costumes to be taken or left as one pleased. Heidegger often recalled attention to the a-historicity proper to the world of technology, which, by reducing everything to *Grund*, comes to lose any *Boden*, that is, any soil capable of giving rise to a true historical newness. But the a-historicity of the technological world, like every element of *Ge-Stell*, probably has a positive valence. *Ge-Stell* introduces us to *Er-eignis* as realm of oscillation above all insofar as it de-stitutes history of its *auctoritas*, making it not a dialectical explanation-justification of the present, and even less a relativistic devaluation of it (which would still be tied to the metaphysical opposition between the value of the eternal and the dis-value of the transitory), but rather the place of a limited coerciveness, of a problematic universality like that of the Kantian judgment of taste.

The Heideggerian meditation on *Ge-Stell* thus takes shape, at least embryonically, as a first direction along the path of an ontology of decline. We may summarize along these lines: (a) *Ge-Stell* lets *Er-eignis* shine forth as realm of oscillation, thus sending us off to rediscover Being not in its metaphysical characteristics but in its "weak" constitution, oscillating *ad infinitum*. (b) To gain access to Being in this weak sense is the only grounding that thought can reach. It is a hermeneutic grounding, both in the sense that it identifies the horizon within which entities come to Being (Kant's transcendental) as a historical-finite thrown project, and in the sense that the oscillation discloses itself precisely as suspension of the coerciveness of the present in relation to tradition, in a retracing that stops at no supposed origin. (c) Retracing *ad infinitum* and oscillation are accessible with a springing that is simultaneously a leap into the *Abgrund* of the mortal constitution of being-there. Or, in other words, the liberating dialogue with *Überlieferung* is the authentic act with which being-there decided for its own death, the "passage" to authenticity of which *Sein und Zeit* speaks. It is insofar as we are mortal that we can enter into, and exit from, the game of transmissions of messages that generations pass to us, and that is the only "image" of Being that we have at our disposal.

There are three elements of Heideggerian legacy in the phrase "ontology of decline" that seem to me essential: the identification of a positive theory of Being characterized as weak with respect to the strong Being of metaphysics, as retracing *ad infinitum* with respect to *Grund*; the identification of hermeneutic grounding as the kind of thought that corresponds to this nonmetaphysical characterization of Being; the peculiar connection of this nonmetaphysical mode of the *Wesen* of Being to the constitutive mortality of being-there.

If one thinks that *Sein und Zeit* began from the necessity of identifying a notion of Being that would permit, above all, thinking the existence of man, historically placed between birth and death, and not just the "objects" of science in their idealized eternity, one can recognize that precisely an ontology of decline responds, finally, to the plan that had been outlined. In the end, it seems that one can sum up Heidegger's thought as the substitution of the idea of Being as life, maturation, birth, and death for the idea of Being as eternity, stability, and strength: that which is permanent *is* not, but that which becomes, in a preeminent way (in the way of the Platonic *ontos on*), is born and dies, *is*. The taking on of this peculiar nihilism is the true putting into effect of the program indicated by the title of *Being and Time*.

TRANSLATED BY BARBARA SPACKMAN

ALDO G. GARGANI
Friction of Thought

The philosophy of culture takes us only so far. The risk that philosophies of culture and philosophies of history run is precisely that of generating intellectual forms that can be overturned into their opposites. This process is carried out in the continuous transmutation in which aspects and conformations of modern culture wish to arrogate to themselves figures that postmodern culture claims as proper and specific to itself. The dogmas of logical empiricism have been dissolved. The rigid program of verificationism, which prescribed translating signifying propositions into statements exclusively concerning sense data, has been confuted, for it based itself on the possibility of an illusory translation insofar as we lack the logical-epistemological foundation for it, that is, the manual for such translation. Referentialistic theories of language have been put into question, since the conditions of truth of the statements cannot determine the individual objects to which those statements refer, with the result that scientific theories are systematically underdetermined with respect to reference.

It is at this point that the cultures of *paradigms*, of the *versions of the world*, and of *conceptual schemes* have been developed. And having reached this point, we encounter difficulties even with these conceptions because one could say that they do not constitute a valid system of self-comprehension of the actual initiatives that men, as subjects, undertake when they describe, refer, narrate, recommend, love, or hate. Perhaps no one would set about telling stories about someone or something, perhaps no one would investigate the particles of physics or social institutions, and no one, finally, would describe the person without whom he presumes he cannot live because he is in love or simply jealous—which comes to the same thing—if all of these acts were considered by their agents as being simply *versions of the world*, *paradigms of description*, or *conceptual schemes*. It is counterintuitive to concern oneself with something whose meaning is in principle proved to be no different than that of adding a further version of the world to the collection of already existing ones. It seems that a friction effect is rendered necessary, a return to the ground of friction in order really to do all of those things that philosophers call versions of the world, paradigms of vision, or conceptual schemes. Men, who are ordinarily the agents, the subjects of all those actions, will feel abandoned and deserted; they will no longer find, that is, the intellectual mirror in which to recognize themselves. What is more, men who carry out that variety of things do not simply expect to see their own movements and their own actions reflected in this mirror. Instead, the mirror is offered to them precisely by the doctrines of versions of the world, of paradigms and of conceptual schemes, a mirror that simply reflects what men do, what they undertake. Does Nelson Goodman, perhaps, not mirror Galileo, Descartes, van Gogh, or Canaletto when he writes that there exists the version of the world of Galileo or Descartes, that of van Gogh or Canaletto? The reality is that what Galileo, Descartes, van Gogh, Canaletto, and all other men as well—who have the honor of being taken into consideration by professional thinkers, by official thinkers—wanted to find in the intellectual mirror was not the mere image of their movements, their actions, of what they did and that consequently takes the name of vision or version of the world.

They were looking for something else in that intellectual mirror: the weight of necessity that had moved them, had made them investigate, write, and paint. They wanted to find the shadow of the gesture of that coercion which—reflected by the intellectual mirror—they were then disposed to recognize and call truth. The philosophical conceptions that limit themselves to characterizing their undertakings and actions as versions, paradigms, conceptual schemes of the world correspond on the surface simply to plausible and possible thoughts. And, in effect, the works by Descartes, Galileo, van Gogh, and Canaletto are in fact defined as possible versions of the world, and it is precisely because they are represented in this way that they prove to be only possible versions of the world. Finally, it is because they are only possible versions of the world that that intellectual mirror is too smooth a surface which leaves one unsatisfied. It is smooth, lacking friction, just as the pure possibility of painting, writing, loving, thrashing, and hating is without friction and smooth. But it is not even the case that men want to smash the mirror in order to see the reality behind it and then measure whether what they have done, painted, hated, or set to music corresponds to it. What they seek is the origin of the necessity that has made them think, hate, write, and paint. In the intellectual mirror they go in search of everything that they have been unable to give themselves and that instead was, or is, burned up as in a collision, with a need that has made them think. Paradoxical as it may prove to be, the state of pure possibility suffocated them, it wearied and rendered them prisoners, and it is only in once again finding the necessity of the origin of their thoughts and actions that they could once more find the truth that is, at the same time, the pleasure of thinking and the deep feeling of freedom.

This truth does not exactly consist in matching a representation with something, with a given, but is a state that is satisfied in once again finding the necessity that was at the origin of an answer of ours. I experience a state of truth not in the adequation of an image or of a representation to something but in the circumstance in which I discover that what I say and do corresponds to the need of saying and doing it, to the intransitive act in the enunciation: in other words, that what we can have written, narrated, loved, hated, and set to music uncovers the destiny of our narration, hate, love, and also of our music. Truth is this interest in the necessity of what is defined as "true." "True" is a primitive term that indicates the acknowledgment of the encounter of a thought with the necessity of this thought. We use the word *true* with regard to something that is said or written, and the word *true* is not a property of what is said or written as, for example, red can be the property of a vase. No. The word *true* is the gesture of recognition and acceptance of the motivation that made one say, write, and speak.

In this way truth is the direction of our interest toward what has made us be all that we are, all that we have said. To encounter truth is for every one of us the meeting of our own destiny in this tangle of possibilities, alternatives to the accidents that are our lives, in which not every possibility, alternative, or accident is the possibility, the alternative, and the accident of our writing and speaking. I can give myself all the possibilities and alternatives in the world. I can slide over the smooth surface of plausible and consistent thoughts for an entire lifetime, but the only thing I cannot give myself is necessity. It will be the acknowledgment of a collision from without that will make me come out from the regime of the indifferent and joint possibilities and that will give me the perception of truth. Only the flash of truth is that which during our life we have not given ourselves but which has come upon us and made us think, and which we recognize as such, which has made us think and recognize (*riconoscere*) rather than know (*conoscere*).

This truth is the echo that comes from within to our thoughts, provoked and kindled in us, we who were strong in possibilities though impotent in necessity, constraint, and still lacking a destiny. Either we go looking for necessity or necessity comes to us: be that as it may, this process is produced everywhere. In mathematics, even if we leave aside the Platonizing assumption of a realm of already constituted mathematical objects and start out from our own free possibilities of postulating and establishing axioms fully independently, we nevertheless end up by constituting numbers that in the final analysis are like bodies that become autonomous, even if generated by an initial act of will: entities, in short, that do not let themselves be indifferently manipulated and that impose constraints on us—on us, who invented them. This is not unlike what happens in writing a novel in which we freely create characters that gradually impose their conditions of necessity to which we, their creators, have to submit. We began by inventing the characters of the book, but at a certain point an astronomic rotation takes place in the pages of the text through which we have *to listen to them* instead of *inventing them*.

What do we make, then, of the statement by Wittgenstein, who, in order to furnish an example of his mathematical constructivism, wrote that if someone were to ask us what number will recur at a certain point in the expression of Greek pi (π), that is, of an unlimited non-recurring decimal number, we could reply in the same way as a writer who said "I haven't decided yet whether I shall marry off the protagonist of my novel"? I would exactly overturn the meaning of Wittgenstein's remark here in the sense that mathematics is not a free construction of our intellect in the same way as a novel is but that the novel faces the same clash with a background of necessity which the mathematician confronts in dealing with numbers. And now the most faithful translation of the analogy would run as follows: I still don't know what will make me think or what will force me to think and to write that the protagonist of my novel must marry a woman. But it is not only a question of knowing, because whatever strategy I brought into play would simply be the strategy of a pure, indifferent possibility. The surface of this ground of possibility is too smooth, and it is necessary to return to friction. The friction, the meeting with necessity, is something that I cannot simply think, and precisely in the sense that it is not that which gives one to think. And when we know, narrate, and speak, we are simple subjects of that case by which that which gives us to think breaks out, and that is precisely the flash of necessity that awaited us from somewhere.

We live, write, and speak in expectation of that accident manifesting itself, which may be a minute event that waits to destine us to think and to make a destiny of ourselves. Only in the meeting with what occurs and makes what we have to think flash as being necessary do we free ourselves from that unendurable regime of simply possible thoughts which is the state of repetition, of tedium, of hysterical rigidity, of the metallic voice in which we do not actually think but contemplate the plausibility of whatever ideation. That regime is one in which everything is in reality similar to everything else, in which everything resembles everything else, in which there is no distance and in which, instead, there is sufficient sterility to keep us from separating ourselves from any thing and from making any decision. Thinking, talking, or narrating without meeting what makes us think corresponds to the reassuring idealization in which the Ego, the subject, presumes or pretends to think *itself, others,* and *the world* and to confer upon them a store of sense. This is the subject that with its colored crayons tints the world that surrounds it. This is the operation that in philosophy has been called the "giving of meaning" or the "attribution of meaning" (the German phenomenologists called it *Sinngenbung*).

If one stops here it is not difficult to believe that one has moved ahead, whereas in reality one has remained at a standstill, affirming that men elaborate or produce versions or visions of the world, that there is the version of the world—as Nelson Goodman writes—by Galileo, by Descartes, by van Gogh and Canaletto. Each of them, each of us, would have his own crayons. The only question that remains would be, What color do you want to make the world today? And yet there has been enormous agitation or, rather, a grand speculative drama has taken place among those philosophers who strove and went to great lengths with "foundational acts of meaning," with "giving of meaning," with those crayons that at times took on the name of transcendental structures of possible experience starting out from a purer, more gaseous, and more rarefied Ego that tinges the world with its lights. The philosophers of this type run about at dawn in the streets of the city in order to color things red and yellow and then declare at midday that the world is red and yellow because of their constitutive acts of meaning. In reality none of them actually went into the city or the countryside at dawn; they stayed in bed like everyone else. Nevertheless, someone did go, but who? Not them, but their delegates, the speculative agents, the "transcendental Ego," the "constitutive subjects of possible experience"—the new modern figures, in other words, of the guardian angel. Certainly, if one can only think, if, that is, one limits oneself only to the initiative of conceiving plausible and coherent thoughts, then one can also think of something like a guardian angel, or of a transcendental Ego that colors the world or that produces a version of the world. Everyone has his or her possible colors, and any color has the value of another.

The subject or the Ego, traditionally understood as the figure itself of "modernity," has been conceived as being a fortress whose consistency depended on the strata of possibility with which it was covered. If, because of some accident, the facade of one possibility collapses, there is immediately another. This Ego expressed itself by way of the possibilities it possessed, but the radius of possibilities upon which the subject is centered is practically infinite. With the stock of its possibilities, of its plausible thoughts, the traditional self-centered Ego or subject can face every situation. Are there not, perhaps, the versions of the world by Descartes, Galileo, van Gogh, and Canaletto—are there not these versions, I say, for this reason? This fortress of possibilities within which the traditional subject is positioned is the factory of plausible thoughts and, more precisely, of thoughts that can arise only from the Ego. It staves off the meeting with what, rather than a thought, rather than a mere conceiving, properly *causes to think* and which is the recognition of that which makes one think. If it does not barricade itself within its possible and indifferent thoughts, the subject ceases to direct its autonomous and independent significations and codifications toward the world; instead, it is a world as occurrence that comes upon, and presents itself before, the subject. And it is exactly at this point that thinking assumes the form of the vicissitude of a destiny that befalls us. From that point on we live in the perspective of an overturning in which it is not we who go toward the world, but instead a mere splinter of reality, an event from the open background of occurrences, which presents itself to us and which demands to be thought. It does not depend on us, and it was not even we who went in search of it. We cannot foresee it, and we cannot even produce it artificially; we can only wait for it and, when it presents itself, observe and listen to it with maximum attention. The so-called free choices of the subject consequently collapse, because what gives us to think chooses to be thought by us. The coercion induced by the necessity we have met frees us from the tedious boredom of the indifferent alternative possibilities with which our old human subject surrounded itself. Paradoxically, that scenario of possibilities, of alternative options, with which our old subject was endowed does not let fly a spark of decision. And that scenario is the theater of the pure, plausible, possible, and conceivable word upon whose stage no flames burn a forest, no sentence causes a war or a revolution, and on which a small movement in the counterweights of the moral scruples of the individual does not subvert the life of a man.

This theater of the subject, self-centered and falling back on possible, coherent, simply conceivable words, is a great exorcism of reality; it keeps reality at a distance, wanting nothing to do with it. It doesn't want the world but *a version of the world*: it doesn't want the event, the accident, but *the conceptual scheme of rational decidability* destined to recognize something that can be defined as an event or is conceivable as an event. But when, and independently of our will or our intelligence, something is given us to think, then we meet the world because the world has presented itself, and from then on it restricts the range of possible thoughts until making our mind the figure of a destiny. This is the thought that truly discovers us, and the paradox of this thought is that it discovers us by coming from without, from an outside of us, but it is as if it had an appointment with a trace of us that we have never known or, perhaps, that we have forgotten, burying it beneath the blanket of possible and plausible thoughts. Truth, then, is no longer the definition of the congruence of a statement with a fact, a sensation, or of the possible agreement of one proposition with another one, but turns out to be the certification of a thought of ours together with the necessity of thinking it.

That I am unable not to think under the aegis of something that has been given as an event is the beginning of every undertaking of truth, as of every true feeling. It is this truth that we encounter that deprives us of or perhaps frees us from all the rest of us which was pure conceivability or intelligent excogitation; and it drains us, it restricts us to being the line of a destiny, of a necessity of thinking. And, finally, this is the truth that gives us to think because our thought comes from without. Hence this motivated thought frees us from that long holiday of life in which the pure syntactic regime of forms has held us, which is a ritual way of existence together like others—like holidays, ceremonies, ornamental practices. Peasants periodically hold a holiday for the grain harvest, but philosophers have held a holiday for thought without there being a harvest to celebrate. Or rather, the holiday was the displacement of a harvest that did not make them *think* but that instead made them pensive. It is also probably for this reason that it was said that every man, the common man, is a philosopher; and all this will even prove to be true if it means to say that every man within himself plays host to the specter of an exorcism, the propensity for movement and psychological substitute formations.

The discovery of one's own self in the form of a stumbling into a thought that comes from outside, that is, of a thought that comes forward in order to be thought, is manifested in what could be defined as auroral situations of nature in which the attitude itself of thought is ramified. In this sense sex also is a scenario and a pulsation of sensibility, and of perception, which are not projections of meaning on someone or something starting out from a subject but which show themselves, take shape and are given shape in a presentation of the other with whom we are involved. Sex is not the thought of sex, nor is it even deposited in another, in an already and previously established form. Sex comes about as the discovery that everyone makes of an important reality of one's own Being by way of an accidental encounter with someone else. Sex is discovered to the extent to which it is encountered, in the sense that sexual experience depends on the body of someone else who, in being lived as otherness, discovers a relation that lay within us in the expectation toward that other person. And all sexual experience is the rotation of this cycle, of this circle of mutual strengthening in which the force of the body of another discovers the force of our own body, of one's own body. We discover our sex for the first time—in reality every time—because the sex of another comes upon us. And it is a not-negligible part of this process that the first sexual relation or relations, or those which in any sense are always the first, prove to be unsatisfying or not as intense as they were conceived to be in the fictions of the simulacra that accompanied us in our sexual fantasies—although, for that matter, no game is immediately intense, full, completely gratifying. The game is not immediately its own intense experience; it does not immediately have the fragrance of unmistakable lived experience. In fact, one must enter into the game, play it, and expose oneself to offense—which it inevitably inflicts because alive and coming from without—in order to contract its bond and be initiated into what will later prove to be its call, its irresistible summons.

The unmistakable sign of these events, which oblige us to think their presentations, is their casual and particular accidental nature, their being, as it were, the indexes of themselves. Theoretical thought, on the other hand, which unravels the threads of thoughts that are simply possible, fatally always sets up great pictures of elevated generality which are, as such, always suspect. The thought of the possible practiced by philosophers carries out a discourse that always takes a world as a totality, as a system of hierarchically ordered relations. But this is a general cosmos that is all held together, without margins or residues, a scene from which nothing and no one advances to meet us. It is a word, a geometrical cosmos in other words, and not reality—which is instead the geometry of a leaf that presumes to say, that reaches out to us in order to say, and that in presuming gives us something to think. Thus a piece, a fragment of reality comes toward us.

If the subject that gave meaning to the world with the simple possibility of its thoughts is an illusion, as is also the practical subject that maintains itself as the total author of its actions, then it is necessary to conceive thinking itself in a completely new way. That is, it is necessary to reconsider this name *thought* for it may turn out to be the name of something entirely different. In the meantime one must begin to realize that *thinking* is not the name of itself, that it is not its own self-designation. It is the name destined to melt away in the scenes of signs, of an entire variety of signs of events that constitute the necessity of an interpretation, but of an interpretation that will not be a version of the world, a simple possible thought of its reality, but rather the bundle itself, the texture of these signs, of their events, their interlacing in an unheard of, unforeseeable commonality. The traditional subject must be demoted or de-strengthened from its position as director of thoughts to become the theater of events. If one can speak of a subject, it is only as a place in which the most disparate events draw together. The unforeseeable ramifications of these events form a commonality only insofar as they coexist; afterward, and only afterward, one wants—given that they are to be found together—to attribute the name of coherence to these events in order to carry out the exorcism of their domestication. The subject now reveals itself as the theater that unites in a single scene the drift of pieces, of figurations and signs which were events. One speaks of them only because they have happened. But the fact that they have happened counts and makes them that great necessity to which the flow of the thoughts of every man is now subject. What one calls the "thought" of men is a communality in which pieces, segments of the scenarios of life, temporal parts, have come to be piled together as events in a communal-ity—which is now the term that has to substitute for that "logicizing" one of coherence.

The word *coherence* corresponded to the illusory idea of a glue that held every step of thought to another in a necessary concatenation, starting from something that was considered the first term of the chain down to the last. Now, however, we know that thought is not a determined sequence between a premise and a conclusion but a scenario looking onto the open backdrop of life and time in the same way in which we say that a room looks onto the garden or that a room faces the garden. Facing this piazza of light which is life and time, we must close our eyes, and certainly not squint, so as not to be blinded and in order to be able to assume a tenable and even indispensable demeanor for our very survival, to be able to relax the features of our face, which would otherwise contort into a grimace under the blinding light. The authority that we possess in the origin of our thoughts is so limited that we may consider our mind as a scene open onto a backdrop of events and times uncontrollable in advance. It is a scene that we have not created but that creates us, and it creates us precisely in the sense that it molds our thought into the form of a theater of accidents and traces that knot themselves together and in which the exorcising rule according to which "if there is a thing, then another thing necessarily follows" no longer holds, but rather where there is one, there is also the other. And the task that inevitably confronts humankind is not the analysis and definition of the ingredients that run together and coexist in that communality, which is thought itself, but the unheard-of effect of that very communality. The unheard-of character of this community that is the theater of human thought is the suspicion or the premonition of a resemblance between ingredients of a scene of thought which, in the ordinary sense of the term, do not resemble each other. The mind of every man is an open-air theater, and it is a place of accidents, casualties that press together in that communality we call thought. There is no preestablished coherency, and every mind is its own diversity of conformation of these signs of accidents and casualities, and it is its own diversity exactly in the same way that every man has his own face and yet is called "man" like so many other millions of men who have their own faces. The same thing is true for the thought of each individual, which is his thought, his own peculiar scene of events and signs held together in a theater that is his mind and that is called "mind" as one calls minds those of other millions of men who have their own minds.

The mind is the community of signs of events, accidents, and traces of scenes of life which do not resemble each other and which, nevertheless, in being together take on a physiognomy due to an interior gesture. Now, the words of our language are the clause of this exceptional resemblance. Words are the intransitive decision, the inaugural initiative that presses together the traits, the most disparate and different signs of events, of the scenes of life. These signs of accidents, these fragments are like pebbles held together in a small bag; each pebble has its own form, its own shape, but they are together; in shaking them together they pass over each other, they are reciprocally modified. And it is precisely this reciprocal modification that goes to make up that physiognomy that we call resemblance and that we define as "unheard of" in that no other pebble is the portrait of another, nor are all pebbles when considered together the mosaic or the picture of the world. The resemblance of this community of traits and signs is the dynamic relation of their being together, of their "doing" together, acting the one upon the other. In order to accelerate the work of the reapers, a ritual was practiced in Mecklenburg in which the name of "corn wolf" (*Kornwolf*) was given to the wolf that presumably lay in wait in the last sheaf of corn, which every reaper thus avoided tying, trying to finish before the others. But "corn wolf" referred not only to the crouching animal in the last sheaf, ready to jump at the throat of the slowest reaper, but also to the last sheaf itself. But there is more: the last reaper, slacking behind all the others, was also called the "corn wolf." As one can easily see, it is a question of signs of different objects and events which, in coming to form part of a single scene, in coming to coexist together, have ended up forming a community of signs which reflects no reality whatever, links up to no external reference, and, in short, is not true with regard to anyone or anything and which, nevertheless, forms a resemblance that is the gesture of an interior and reciprocal attraction of the factors at work. The name "corn wolf" is the clause of this sign community of things which are different from each other, which do not resemble each other, but which coexist together, the one working upon the other, proving at the end to be the gesture of an interior resemblance, an internal relation that is intransitive insofar as it is properly inaugural. And it is inaugural because it does not need to mirror or conform to some previously established agreement or identity.

The resemblance of the community of ingredients of a man's life, constituting the physiognomy of his existence and mind, is not the icon or reflecting image of some piece of reality: that resemblance consists in the echo that each fragment of the life of a man reverberates in the direction of other persons. The life of every man is actually the context of a disorder—what life is ordered?—but a disorder whose components and constituent elements have set to echoing each other, thereby forming—and not reflecting—a tracing of internal relations and summonses. The word, the name, *designates*, it establishes its power of designation in that it exiles itself as fulfillment of this community of factors which coexist together—where if there is the one there is the other, where the one gives the other in an echoing thought.

For Marcel Proust, "Balbec" is the name of a community of signs; it is the context of an interweaving of graphic and phonematic features that model the curves of a scene effluent with meaning ("una scena influente di senso"). With its pointed, cutting graphemes and the phonematic traits of the broad sound of the first syllable, which is forced into a brusque and closed finale, the name "Balbec" in the *Recherche du temps perdu* echoes at the same time the broad wave of the Sea of Brittany with the ample caress of its seduction, and the spires and steep profiles of a Gothic cathedral. "Balbec" is the name of a communitary relation of echoing physiognomies centered upon a passion that, even prior to being the passion of a cognitive and desiring subject, is the very passion that each of these traits takes toward the others; it is all the love of the one for the other, the exchange of their mutual, reciprocal, and sliding caresses. The word radiates from this context of signs, of acoustic and graphic seeds; it is their community, their reciprocal passion. This name, "Balbec," is the theater of a life, the effluent scenario that actually comes to coexist in a man, who only then we shall recognize as being equipped with what we call "consciousness," "subjectivity." Visual and acoustic traits, sounds, scenes, a rain of signs, have engraved an influential scene; sufficiently influential as to make an eye a gaze and a neurophysical apparatus a consciousness.

There is no thinking nor even a *desiring*, or wanting, that is a separate act or process, autonomous with respect to these effluent scenes of signs which take place, which coexist together, and whose final formula cannot be foreseen, or even produced artificially, but which must only be awaited and listened to. Self-conscious reflecting, subjectivity, is only the terminal point, the extreme point that surfaces from the theater of scenes, signs, accidents, and events which—as it were—aggregate *in us but prior to us* and which are all the passion of our living. Consciousness is the theater in which one ascertains, not that a process has been carried out, but that a process has been carried out and now reaches us: this is consciousness together with the backdrop of necessity which is the imposition of its own thought. This relation of internal resonances, of echoing signs that come to coexist in a community that is the antecedent of thought and that constitutes the reference of names we use, receives the definition of *coherence* when it is transposed onto the register of the "logicizing" discourse that classifies verbal materials without plumbing their internal seductions—that is, losing their passion. The strabismus of the gaze, from which one uses the term *coherence* in the usual "logicizing" sense, consists in assuming that community or context or coexistence of echoing signs, of passional seeds, to be the fiber of a causal concatenation that would sustain the unity of thought, inexorably binding the steps in which it is carried out. In this strabismic gaze coherence is thought of as a cord that passes through and ties the steps of a thought.

But in fact there is no cord of this kind to constitute the bond of thought. For that matter, not even in its ordinary and material meaning does the cord exist as conceived by the "logicizing" philosophers, for the strength of the cord consists in the superimposition of one fiber over another, and if we unwound the fibers we would have no cord. Analogously, if we were to remove the signs, events, the different and heterogeneous pieces that from the open backdrop of life have come to crowd together in the community or coexistence of the mind of a man, as, for example, the idea of "corn wolf," there would be nothing left to think—in fact, there would no longer even be thought itself. Thought, then, is not a specific, autonomous, and independent essence that interprets, that forms ideas, versions, or conceptions of the world. And equally little can one say that thought generates thoughts or that the will forms voluntary acts. No, thought is the concrete weaving formed from an unpredictable constellation of events, of diverse scenes, of signs of passions; and if these same events, scenes, these different signs, details introduced by chance or by the accidentals of life, by the fragments that each person lives of this existence, were removed, there would no longer remain anything that we could call "thought." Thinking is the state of the coexistence of these traits, of these signs of events, of these figurations of a scene of life, of the signs of their passions. In this sense, thought taken in itself no longer exists. And if one insists in taking it on as an independent and autonomous existence, in accordance with the register of "logicizing" philosophy, it is only a misleading duplication. Thought cannot distinguish itself from passions and emotions established by an effluent scene of signs, of seducing figurations, of graphic and acoustic seeds which form a coexistence or a community or a context in the mind of a man and which are his entire destiny.

Thought is not a specific essence, not the autonomous activity of a consciousness, but the state of a variety of traces, of accidents that coexist together; it is the relation of their emotions and passions; it is the same perturbation of their presenting themselves, of the fact that they take place. If thought per se, as an independent essence, is an illusion, an idealization like that of its correlated concept of subject as substratum and director of thoughts and actions, then so too is the notion of man endowed with value insofar as he is the so-called desirable man: that is, insofar as he conforms to some ethical norm or principal of consciousness to which he can be referred back. If man is a point of intersection and of listening to events, signs, of a figural conformation of graphemic and phonemic traits, if his thought and will are only the state or context of these signs which are recalled as passions of an echo that rings in him, there is then no thought that is independent of these concrete traces, of these influential scenes of his life, just as there is no independent and autonomous will with respect to the things that are wanted, to the decisions that are intransitively established.

All of this implies a passage from the consideration of man as a desiring and desirable Being, capable of being evaluated according to preestablished norms and criteria, to the state of the real value of man, *for what he is*, and not simply on the basis of his desirability. I mean that in the rotation of the axis of these considerations the value of man is acknowledged not in the will within him, or in that of another subject which directs itself onto him, but instead in what is willed in him, in what within him is not *conformity to a will* but a completed realization, perfection of the exactness achieved. Value consists not in the *will* or *desirability*, as was traditionally maintained and as actions—and not by chance—were traditionally evaluated, but in the completeness of an intransitive decision which is comparable to an effluent scene, which can be read and or seen as a situation of nature, which is self-willed instead of conforming to the criterion of any desirability and which, in consequence, has been arrived at as nature, which has imposed itself like a tree, a boulder, or the profile of a promontory. Value does not consist in the will to value or in its desirability but in what actually turns out to be willed, which has actually been registered, which has proved to be an effluent scene to see and listen to.

In short, it consists in that which has been translated into necessity and which now stands before us as a nature. Like a nature, it does not correspond to the concepts and attitudes of "good," "evil," "beautiful," and "ugly"—in the event we even had these concepts prior to meeting the trace, the sign that those concepts, by way of their completeness and exactness, also express necessity, that necessity which gives to think rather that *the necessity which thinks*, attributed by philosophic tradition to the so-called subject, to the transcendental Ego, to that Ego which carries out the giving of meaning (*Sinngebung*). Value is the fulfillment, the self-fulfillment, like that of trees which remain still and well rooted while they are lashed at by the wind and other elements. Value is the fulfillment of a nature, in the same way it is equally the completion of the character of Singleton—regarding whom, in answer to the objection raised by John Galsworthy, who did not agree with the value attributed in *The Nigger of the Narcissus* to this character of a modest and ordinary sailor, Joseph Conrad observed that Singleton, on the contrary, was a highly significant and important figure because he lived in harmony with himself, because he was simply his own nature. And nature here is no longer that of the cosmos transfigured by the providential design of a thinking subject, giver of meaning and bearer of values, but the nature that is its own fulfilled will—as happens when one goes out into a city's streets at dawn, where there is no longer trace of good or evil, of right or wrong, but where everything is simply its own force and strength.

The traditional philosophic subject, bearer of thought and will as independent and autonomous essences, is, in reality, *the radical negation of every thought and of every will*. Precisely because of the circumstance of professing to think the world from the distance taken by a thought in believing itself self-sufficient and self-centered, that subject does not know how to confront the world; it is not capable of tolerating the necessity with which it gives itself to think and the effort that this imposes on the human mind. That subject does not have the courage to listen to the world; it is incapable of assuming a self-control tenable in the face of the storming of the perturbing voices that begin to echo, surrounding it from every side and tugging it along a thousand paths. And given that it does not possess the courage for all this and is incapable of truth, that subject gave and gives itself over to constructing versions of the world, conceptual schemes of ordering and classification as *substitutive psychological formations* of the act which demanded the meeting with necessity, the listening to the perturbing signs which we do not seek but which present themselves "in person" so as to come into collision with us. That subject is the figuration and idealization of the impotence of he who does not know how to place himself as a completed and determined will, as perfect as a nature, and that also in this case produces a substitutive formation by way of the theoretical recommendation of desirable values, of norms and metrics of trustworthy and guaranteed action.

In both cases this subject of traditional philosophy, this common man who is the child of our civilization, is he who, in having the courage neither to know, act, tolerate, nor to suffer, carries out *a psychological operation of displacement*, giving other men lessons concerning the ideal and founded conditions of knowledge and norms regarding the desirable values of practical conduct. Given that it is not the completed nature of any of these things, this subject barricades itself within itself; it raves about theoretical machinations and speculative regulations with regard to a world it is afraid of nearing. The reign of Kant ends, the ideal community of communication in which every interpreter, as self-centered subject in the full transparency of itself to itself and in the revealed presence of Being, is the contemporary of every other equally self-centered and self-conscious subject in the presence of full and actual Being: these and other similar theories (Schleiermacher, Dilthey, Apel) are the displacements and the substitutive formations by which a long philosophical tradition has plotted the narcissistic isolation of the Ego. They are the deviating idealization of a thought that chooses the hysterical strategy of self-reference in order to avoid being invested by the passions in objects, that still chooses to lay down the law according to a register of austerity rather than listen to objects, signs, the circumstantial events opened up by the world's horizon, by the scenes that come to form that unmistakable interweaving of necessity and pleasure which is what actually gives men to think.

It is necessary to come out into the open air of the scenes of the world and of the most incredible interlacings of that varied impasto of events in order to trace once again the matrix of history according to which we, above all, occur. Even the most linear human biographies are the result of the most disparate ingredients, of factors that among themselves are extraneous and divergent and whose only connection is not that of the analytical rule of coherence but that of what one can only define as a destiny. There is no linearity in any human biography but only the incursion of events, signs, figurations, and most disparate repertories of life, which are condensed in the trace of divisions, distances, and conflicts.

It is because of this disparate and diffused origin of the destiny of each and every man that the word *conflict* occurs in the transcription of every life and that the elaboration of the life of each individual is a nature that looks onto a backdrop of fear. The community or the context of the pieces of scenes and signs of reality which form the mind and destiny of each man breaks away from the difference every time; it is intrinsically marked by the role of residues that have remained outside, by everything of which there remains not the presence but the trace. It is perhaps for this reason that the biography, the mind of every man is involved in a destiny that is light and darkness; apprehension, appropriation, and loss; revelation and concealment. It is upon this background that things both are and are not, that there is transparency and opacity, and that every flash of truth and of meaning accepted proves to be surrounded by something vaster. And from the fullness of the meaning of a circumstance which is certified as being promising, full, and imminent, we are nevertheless driven toward its edge where it borders on something inscrutable which strangely, even if by definition it does not belong to us, seems precisely to contain the mystery of ourselves or, at least, the interrogation regarding ourselves and the circumstance we live although we are not completely ourselves. This very possibility concerning a situation or a moment—even a fleeting second—of our life that contents us, that makes us happy, nevertheless (in fact, perhaps precisely for this reason) tends toward that shadowy zone, in the direction of the unknown border or margin. And the tension then rises—on the basis of the present happiness—to cover also that unexplored region which is like the unconcluded and indefinite heavens of our destiny. We are also terribly interested— and we must say it—in that zone of the heavens and that horizon which encircles us from afar. We are made up of extraneousness, and extraneousnesses above all interest us. We could not decline the suggestion of again and anew wanting to make ourselves by way of that diversity and those clashes which are also already the amalgam of our past.

TRANSLATED BY HOWARD RODGER MACLEAN

MARIO PERNIOLA
Venusian Charme

I. Seduction, Love, "Charme"

It would seem at first glance an impossible undertaking to break the link between erotism and evil without restoring the illusory, transfigurative positivity of love. The myth of Don Juan has established and maintained such a link between the sexual drive and the negative, between the dynamics of pleasure and sin, for at least three centuries. Though recent attempts to reconsider erotic life through the concepts of *seduction*[1] and *love*[2] move in opposite directions, they do converge on one point: both discourage the search for a path that would be independent of either the libertine or the romantic tradition. Both react energetically to the banalization and loss of meaning of sexuality in contemporary society by rethinking, in original and subtle ways, the two fundamental concepts through which the West has given meaning to its erotic life. But precisely for this reason, and notwithstanding the modifications they bring to the notions of love and seduction, they remain within a tradition that contemporary society would seem to have discarded. One may take the subjective aspect away from seduction and submit it to the rules of the game, but it remains always challenge and negation. One can make love more anarchic and disordered by multiplying its manifestations to infinity, but it will always tend toward transcendence. Of course, it is characteristic of our times to be beyond good and evil, to have little tolerance for truly immoral or truly moral behavior, and to overturn either into its opposite, and, finally, to sink both into a state of indistinction in which everything is reversible into everything else, everything is confounded with everything.

In the erotic civilization of the past two centuries, seduction and love are complementary dimensions that describe, respectively, the most common masculine and feminine behavior. For every Don Juan who seduces, there is a Lady Anne (or more) who loves him. Of course, one can significantly modify this paradigm by inverting the roles. One may say that seduction, as a strategy of appearances, is first and foremost feminine. The feminine would not be that which is opposed to the masculine, but rather that which seduces the masculine. By the same token, one can identify the solution to the current crisis of masculine sexuality in an amorous disorder in which masculine erotism can, by abandoning the code of virility, open itself to an emotional intensity previously unknown to it. Both orientations tend toward going beyond the distinction between masculine and feminine toward *transexuality*. But both, precisely because they remain prisoner of the notions of seduction and of love, can at best overturn traditional attributes without succeeding in going beyond the erotic civilization that created the myth of Don Juan and is the apologist for the redeeming power of feminine love. The heart of the matter is not sexual but philosophical: the waning of masculinity and femininity depends upon the dissolution of the concepts of seduction and love, and on the search for an erotics independent of both libertine negation and romantic transcendence.

This new erotics must therefore stand upon notions independent of a prejudicial critique or metaphysics. Better than the word "fascination," too connected to the enthralling magic of the gaze and its malevolent powers, the word "charme" presents itself as open to various uses and suited to indicate divine emotions as well as sexual attraction.[3] Such polyvalence acquires better definition when placed in relation to the impersonal notion of "venus" as it was understood in archaic Roman religion, before it came to name the goddess and was confused with the Greek Aphrodite. The interest that the archaic idea of "venus" excites today is not due to a generic contemporaneity of that which seems least contemporaneous, but to specific reasons connected to historical research and contemporary experience. In fact, historical research can show such a notion was not dissolved by the Hellenization of Roman religion, but remained alive and active in more or less subterranean forms in the West. At the close of the erotic civilization dominated by the figure of Don Juan and romantic love, the idea of a "venusian *charme*" free of mythological baggage reemerges. It is articulated by means of an analysis of the four fundamental words Robert Schilling has inferred from the linguistic study of the term "venus": *veneratio, venia, venerium,* and *venenum*.[4]

II. "Venus" as Veneration

If seduction is challenge, transgression, and negation, venusian *charme* implies an opposite attitude of acceptance of the given and affirmation of the present. This does not mean resigned and forced acceptance, *obtorto collo*, as seems implicit in the verb *colere*. Nor does it indicate good-natured consent, as in *placare*, but rather full assent, a disposition of the will to *say yes*, to venerate, to give oneself without reservation. Raymond Radiguet, one of the most important twentieth-century interpreters of venusian *charme*, has captured the essence of *veneratio*: "It means to devalue things and misrecognize them, to want them to be different from what they are, even when one wants them to be more beautiful."[5]

Veneratio is a silent movement because it suspends and silences the subjective desires, individual passions, and disordered affections that would impose themselves noisily against the divine, worldly, and human givens that require their realization without seeing or understanding reality, and that rush toward utopia and destruction, oscillating between arrogance and desolation, exaltation and depression. The Roman goddess Angerona, goddess of will and occasion, seems to personify the silent premise of all veneration: her simulacrum held a finger to her lips, ordering silence.[6]

Veneratio means to say yes above all to the gods and hence to abandon totally all Prometheism, all *hubris* in the face of the divine. Man must please the gods; they must be enchanted, enthralled, fascinated by whoever turns to them. The *captatio benevolentiae* is the starting point of this erotics. But the gods must be silent if they are to be venerated.

It seems that the Romans introduced veneration at precisely the same moment that they took speech from the gods, deprived them of myth and the narration of their feats. Georges Dumézil has shown that the gods of the Roman religion are the same as those of the Indo-European pantheon, but demythified, silent. Unlike Etruscan religion, Roman religion has no revelation: the Sibylline Books are a mere collection of rites to expiate the prodigal. The injunction *favete linguis* that invited participants to facilitate the ceremony's course with silence was therefore addressed to the gods themselves.

Veneratio means to say yes to the world and hence to abandon resentful attitudes, preconceived criticism, or systematic refusal of the present. It is impossible to be *charmant* if one is not at peace with the world, with the spirit of one's time, with one's surroundings. To venerate Venus in the world means to be willing to recognize the variety of her manifestations and to will them according to the occasion. Chastity and orgy, marriage and prostitution, monogamy and polygamy, homosexuality and heterosexuality: these are not incompatible realities among which one must choose once and for all, but situations one may appreciate in the proper moment. Yet the condition of their appreciation remains their silence, their discretion, their demythification. To be *charmant* means not only to be ready for the opposite with the same indifference, but also to maintain a detachment that allows one to respect the cadence and rhythm even in the most decisive action. Venus presented herself to the veneration of the Romans in two apparently incompatible forms: as Venus Verticordia and as Venus Erucina. The cult of the former was aimed at turning the minds of young girls and women to chastity. The cult of the latter, of Sicilian origin but promoted to the rank of Roman divinity and honored with the erection of a temple on the Campidoglio, was instead closely linked to the practice of prostitution. The attribution of such opposing qualities to the same goddess does not arise from a nihilistic attitude that wishes not to compromise itself and hence favors one quality at one moment and another at the next, but rather from a profound intuition that manifests itself in the quality of the cult. Diodorus Siculus recounts that when the Roman magistrates traveled to Sicily, they always honored the sanctuary of Eryx with sacrifices and homages and "in order to please the goddess, they forgot the gravity of their mission in order to make merry in the company of women."[7] These magistrates were thus *charmants* in the eyes of the goddess before they appeared to those of her priestesses precisely because they took a detached interest in pleasure, a nonparticipatory participation. The poet Giambattista Marino astutely captured this venusian indifference in regard to chastity and lust when he shows in his *Adone* "Venus applauds obscene works no less than their opposite."[8]

Finally, *veneratio* means to say yes to oneself. Not, of course, to one's own desires, dreams, and ideals: all these things are too imbued with negation and absence, too abstract and inconsistent to be truly retained as elements or aspects of oneself. Seduction may be rightly defined as a magic of absence,[9] but "venus" is, quite to the contrary, inseparable from presence, from one's own situation, from that which is given to us. To venerate means to be at peace with oneself, to know how to will backward, to want that which has happened, to transform (as Nietzsche's Zarathustra says) every "so it was" into a "thus I willed it to be." Veneration is *amor fati*, a will to want that which has been and is, yet not in order to remain locked within the circle of an eternal return of the same, but on the contrary in order to want the present without being conditioned by its contents. It is thus the opposite of quietism that abandons itself completely to fate. It is the human participation in veneration that transforms any event into destiny, because the entire past was already "destinal."

And yet the repetition and devotion implicit in *veneratio* are not a true faithfulness. By silencing the gods, the world, and oneself, veneration is the premise of a mimeticism that distorts all the more the more formally identical it is to its model. Radiguet remarks, "Nothing resembles things themselves less than those things which are close to them."[10] This is especially evident in the consequences of the ritual of *evocatio*, used by the Romans to invite the enemy's gods to leave their cities of origin and come to Rome. The formula used to "evoke" foreign gods was "veneror veniamque peto." It is evident that the veneration of foreign gods required the initiation of a Roman rite dedicated to them, a rite that was more dislocation and distortion, *déplacement* and *détournement*, than respectful procedure. At the base of Roman religious syncretism and of its extraordinary ability to assimilate the most diverse cults, one finds an attitude of veneration and acceptance that is not mere affability, but rather a most original erotic strategy, subtle philosophical and political thought. It would be a grave error to consider veneration a weakness or meekness; it is rather the arm of a *pium bellum*, of a good war conducted without resentment. The association of Venus and Mars that the Romans probably borrowed from the Greek couple Aphrodite-Ares therefore reveals a meaning that is deeper and more exquisitely Roman. The connection between veneration and war figures also in *devotio*, the rite in which a commander in particularly dire straits recited a formula, a *carmen* that dedicated him to the Manes and to the earth, in order to obtain victory.

His offering himself to the beyond reveals a relation between venusian *charme* and death that is radically different from that which links Don Juan to the statue of the "commendatore" in seduction, or that which links Tristan to suffering and catastrophe in love. Whereas Don Juan is forced to accept the statue's fatal invitation,[11] and Tristan's love is by definition opposed to mundane reality,[12] the Roman commander spontaneously consecrates himself to death in order to win. For him, being among the Manes is once again a way to say yes to the present.

III. "Venus" as *Venia*

If *hubris*, the arrogance implicit in seduction, invites *hate* and punishment; if amorous suffering is compensated by moral redemption and spiritual salvation; the *veneratio* of venusian *charme* solicits *venia*: the benevolence and grace of the gods, the world, and man. *Venia* is not properly speaking forgiveness, because no sin or even indulgence has been committed. Nor is it an allowance of space and time for repentance, since no deviation or error has occurred; in the venusian dimension, man is innocent. Of course his innocence is not ingenuous, spontaneous, and natural; it is an innocence located beyond good and evil because *veneratio* initiates a new beginning. Titus Livy tells that after the *devotio* of the consul Decius Mus, the Romans "took up the battle as though the sign had been given for the very first time."[13]

A conspicuous part of the *charme* that the venusian perspective has exercised upon poets in particular derives from its character as repetition that presents itself as different, other, not identical to the preceding one, to the model or original. Here we find a explanation of the link between Venus and spring that is less banal than the usual generic reference to "enchantment and the flowering of nature."[14] The return of spring is enchanting because it initiates a transition, a passage from the same to the same. The refrain of the poem *Pervigilium Veneris* brings to the fore the cancellation of experience, the indifference in the face of past erotic experience: "Cras amet qui numquam amavit, quique amavit cras amet" ("Let those who have never loved love tomorrow, let those love tomorrow who have loved").

Venia is the consenting response of the divinity who has been an object of veneration. In the mutual relation of *veneratio-venia* that is established between man and divinity, Venus combines in herself the two poles of the relation: she says yes to those who, inspired by her, have already said yes. She is thus the propitiator par excellence: she suggests *obsequium* and is *obsequens*, is propitious and compliant to whoever already moves within a horizon of propitiation and condescension. Roman deities are endowed with *venia*, and Venus is by definition *obsequens* because assent and affirmation are implicit in the very notion of *numen*, of divine power. *Numen* comes from *nuo*, to nod. Of course this does not mean that the gods may not be irate or hostile at times, but there is always an expiatory or propitiatory rite that reestablishes the *pax deorum*. It is this faith in the fundamentally favorable nature of the divine and of the present that allows the Romans to deify (to the horror of Augustine and Hegel) even the most harmful forces like fever and the goddess Lua, symbol of disorder and destruction, as well as the most secondary and laughable forces like those named in the Indigitamenta, because all these participate in some way in presence. Upon this faith is founded the possibility of assimilating the most diverse religions to that tolerant syncretism of the strangest cults that characterizes the development of Roman religion. They only thing that is truly unassimilable to the Roman pantheon is moral radicalism, precisely because it negates the present in the name of an "ought to be," of Sollen, of utopia.

The concept of aid is implicit in venia. It is curious that the verb nuo (I assent) is confused with an archaic nuo that means "I suckle, I nurse" (whence nutrix). The idea of benevolence and of venia thus seems linked to that of aid given in early infancy, in a state of extreme need. No matter how much this may tempt us to consider Venus as one of the many manifestations of the Mediterranean archetype of the Great Mother, such an identification would overlook the essential point. Readers of the Aeneid will certainly remember the episode in book 12 when Venus Genetrix runs to the aid of her son, Aeneas, who has been wounded in the battle against Turnus. Venusian literature is equally rich in examples that intend the aid of Venus in an erotic sense, from the Camoens of the Lusiadi (for whom Venus conjures up from the sea a lovely island inhabited by quite compliant nymphs who give themselves in the most voluptuous ways) to Radiguet, for whom Venus ironically "lets us glimpse her secrets, her fruits" unconsciously in sleep.[15] But the notion of aid implicit in venia is much broader than that of maternity or sexual surrender: it must be understood in all its material and spiritual latitude. Venus is obsequens not only like a mother who nurses or matrons who, fined for their adultery, financed the erection of Venus's first temple in Rome in 295 B.C. The characteristic of her venia is of the philosophical order: it implies above all a willingness more general and vast.

If veneratio is to say yes to the gods, the world, and oneself, first silently and then according to ritual carmina, venia is to receive a yes from the gods, the world, and oneself, at first through a mute nod, a sign of approval, an intimate consent, and then through a word that is almost "independent of him who speaks it" which means "not insofar as it signifies, but insofar as it exists." This is the meaning that Emile Benveniste attributes to the root *bha—whence for (to speak) and its derivations fas, fama, and fabula.[16] Of course the idea of fas understood as a divine word in a mute pantheon presents some difficulty, but the important thing is to point out the affirmative character implicit in the word fas and its ritual, demythified aspect.[17] Thus the term fama seems to have originally had an affirmative intention. Finally fabula, the fabulation of oneself, may create a persona (in the Roman sense of mask) but not a subject: the doubt about its reliability from the very start prevents the individual from failing *pietas* and becoming arrogant.

Just as *veneratio*, the giving of praise, turns into a mimeticism that dissolves the meaning of that which it praises, so *venia*, the receiving of praise, finally annuls the content of that which is praised. The facility with which one is accepted as a sexual partner in contemporary life is part of the venusian *charme*, but this does not justify any particular complacency nor does it authorize any intimacy. These encounters, consummated without pathos and without anyone attributing any particular importance to them, have a profound enchantment: they are appreciable ceremonies precisely because they are empty. They are under the sign of Venus: the *venia* exercised in them annuls all vanity.

IV. *"Venus" as* Venerium

The luckiest throw in the game of dice, obtained when the four die each showed a different number, was called *venerium* by the Romans. This illustrates the relation between Venus and success. While seduction seems connected to an unhappy destiny,[18] and love reciprocated has been wittily defined by Beckett as a short circuit,[19] venusian *charme* is inseparable from success and a happy ending. Thus to remain locked within the metaphor furnished by the game of dice is misleading: Venus has nothing to do with chance. Her protégé would be like a player who "executing 100 throws, 100 times gets the *venerium*,"[20] but for the Romans such pretension would be an expression of the arrogance that is precisely the opposite of the venusian spirit.

Presumptuousness—Livy calls it *iactantia*[21]—was the sin of the inhabitants of Praeneste who believed they could always win because they were protected by Fortuna Primigenia, who is foreign to the spirit of the Roman religion.[22] Fortune, mere chance, does not at all occupy an eminent position in the Roman religious cosmos, and the idea of an essential and absolute originality is opposed to the experience of a city that was born and developed through assimilating and distorting mechanisms.

It is not by chance, then, that sources exhibit traces of a polemical attitude on the part of the Romans with respect to the Praenestine cult of Fortuna, an attitude apparent in the prohibition on consulting its oracle. The Roman suspicion of the concept of fortune has a philosophical basis: it depends upon the contrast between a voluble and uncertain *fortuna* and the venusian *felicitas*, "solid and sincere."[23] That Servius Tullius, son of a slave and patron of slaves, fortunately conceived and made king, had according to tradition dedicated a temple to Fortune tallies perfectly with this assertion. As Angelo Brelich observes, Fortuna in Rome is the goddess of slaves and those who live by their wits ("sine arte aliqua"), of those whose only remaining hope is for a stroke of luck. The goddess Spes is in fact associated with Fortuna in the Praenestine sanctuary.

The success of Venus's protégé is not due to aleatory factors, for he is not under the sign of hope, which awaits events that may or may not happen. Nor must he be tainted by arrogance, and hence does not depend upon the presumption that certain favorable events necessarily occur. *Felicitas* consists in considering whatever happens to be favorable. Sulla, to whom the cult of Venus Felix is attributed, seems to have cultivated this idea implicit in the notion of venusian *charme*. He seemed to attribute greater value to his own image as *felix* than to real political power and in any case attributed the latter to former. According to Plutarch, he maintained this opinion of himself to the very end, in spite of suffering from a horrible intestinal ulcer that destroyed his flesh, transforming it into lice and dirtying him with an unarrestable flow of rotten matter. Despite this infirmity, which forced him to immerse himself in water several times a day with no results whatsoever, he never ceased to consider himself *felix*. Two days before his death he ended his memoirs, asserting that "after he had led a life of honour, he should conclude it in fullness of prosperity."[24]

By associating the concept of *felicitas* with that of *victoria* and inaugurating cults and temples dedicated to this new goddess, Pompey also put himself under the protection of a Venus Victrix. Such a choice did not prove a felicitous one, since it conflicted with Caesar, who placed Venus in person among his ancestors! Appianus recounts that the night before the battle of Pharsalus, Pompey dreamed of decorating the temple of Venus amid the applause of the people. Awakened suddenly, he realized that the dream was not in his favor and, profoundly unsettled, went toward defeat substituting the battle cry "Venus Victrix" with "Hercules Invictus."[25] The episode demonstrates that venusian *charme* is not reducible to a hope for a military victory: it transcends the good or bad outcome of a single conflict. It is not success in itself that makes one *charmant*, but *charme* that predisposes one for success. The very concept of success loses its objective characteristics in the venusian perspective and becomes an attribute of enchantment: the Romans knew quite well that there were victories that were worse than a defeat, and defeats more providential than a victory. Caesar's decision to erect a temple not to Venus Victrix, who had helped him at the Battle of Pharsalus, but rather to Venus Genetrix is illuminating: he considered victory merely a consequence of venusian protection.[26]

V. "Venus" as Venom

The word *venenum*, like the corresponding Greek term *pharmakon*, presents a double meaning, for it can be used both positively and negatively; it thus originally seems to have indicated the power of venusian *charme* in its multiple manifestations.

This affinity with the Greek term does not, however, illuminate its conceptual dimension, which is essentially Latin and is determined in opposition to the horizon opened by the noun *pharmakos*, related to *pharmakon*. In Greece, the scapegoat sacrificed (put to death or expelled) in order to purify the city of the ills that afflicted it was called a *pharmakos*. To this end, a certain number of degraded and useless individuals were regularly maintained in Athens at the state's expense.[27] René Girard sees in this custom a manifestation of sacrifice whose essence consists in the exercise of a ritualized violence that purifies and guards the community from the spread of unrestrained and total violence. This theory is founded on the presupposition that only the ritual repetition of violence, by provoking a cathartic and beneficent effect, can distance and preserve a society from barbarism. Human or animal sacrifice (implying bloodshed) is the only *pharmakon*-remedy to the *pharmakon*-venom of generalized violence: "non-violence appears to be a gift of violence".[28] As Derrida has shown, this perspective remains operant within Greek philosophy, in particular within Platonic philosophy.[29]

Though there are a few sporadic cases of human sacrifice and ritual expulsion from the city to be found in the religious history of Rome, the word *venenum* turns our inquiry in a different direction. "Veteres vinum venenum vocabant," says Isidorus of Seville. This evidence, together with the study of the Roman feast of Vinalia, points out not only the sacred character of wine understood as the venusian drink par excellence,[30] but also the meaning of the substitution of wine for blood in sacrifices. The sacralization of wine in Venus's religion plays a role completely different from the one it plays in Dionysus's religion: in the most ancient Dionysian tradition, there is no reference to wine and the relation between the two is only established retroactively.[31] The Dionysian intoxication comes from the homicidal fury of the *sparagmos*, the tearing to peices of the victim, consumption of his blood and flesh.[32] The bloody sacrifice of Dionysism is the *pharmakon* that restores peace and social order. In the religion of Venus, however, the *vinum-venenum*, significantly considered the "blood of the earth," immediately takes the place of human blood and implies a refusal of violence even in its therapeutic and prophylactic uses. That the *pax deorum* is reestablished by means of the libation of the contents of the grape-harvest jars, rather than by means of bloody sacrifices, is a fact of enormous anthropological importance. Venusian *charme* thus locates itself at the antipodes of orgiastic intoxication. While the attraction exercised by Dionysus derives from the ritualized and controlled imitation of an originary and founding violence, the attraction exercised by Venus is, on the contrary, connected to a sort of displacement, *déplacement*, transfer: by offering wine rather than blood, Venus establishes an astute mimeticism that exalts the grace of *détournements*. *Venenum* also means dye, tint, color, and by extension makeup, *maquillage*. In this way the cult of Venus interprets a profoundly rooted orientation in the Roman spirit, traditionally attributed to the second king of Rome, Numa Pompilius: in response to Jove's request for human sacrifices, Numa did not refuse but displaced the meanings of words by offering him heads of onions rather than human heads, hair and pilchards rather than men.[33] It is significant that Jove appreciated Numa's translation, in contrast to the Greek Zeus, who (as Hesiod recounts) did not forgive Prometheus for having given him bones covered with fat rather than flesh as a sacrifice. Also in this perspective is the tale of a certain Papirius, who, in an era when it was customary to promise entire temples to the gods as a vow, promised Jove a "pocillum mulsi," a glass of honeyed wine, and obtained complete fulfillment of his requests.[34]

Venusian *charme* is certainly linked to appearance, but not necessarily to "beautiful" appearance. The existence of a cult devoted to Venus Calva, whatever its origin, is yet further evidence of a religious disposition oriented toward an innocent *déplacement* that excites, not the wrath of the gods, but their smile. Demythification is also dedramatization: exaggerations and fanaticism are alien to Roman religion, which rejects the absolutist claims implicit in the delirious experiences of Dionysism.[35] Dionysus's religion knows ecstatic joy but has none of that humor—benevolent and astute, prosaic and witty—that is an essential part of venusian *charme*. The poets have been the interpreters of this aspect, from the incomparable Giorgio Baffo (whom Apollinaire considered the greatest erotic poet of all time) to Radiguet. Baffo's Venus, who "sprawled out on the grass in a delightful garden with her lover" teases her companion with these words: "Come on, then, my lovely, give me the precious juice of your blessed prick, for I prize the juice of your little dick more than muscatel" and concludes: "May those who don't fuck go to hell and become so many marmots. But let the first to have screwed be praised, honored, and crowned," belongs to the same erotic intuition that gives rise to the Bald Venus and *vinum-venenum*.[36] The demythification that exchanges wine for blood in sacrifices and onion heads for human heads is nonetheless not mere banalization or triviality: disenchantment does not eliminate enchantment, and exteriorization maintains a purity of its own. Venusian *charme* does not arise from a dialectic of concealment and unveiling: it presupposes an already uncovered and available reality. Enchantment does not depend upon what is hidden or revealed, but on the transformation undergone by the "crudest" and "most obscene" reality. If there is still a secret to be revealed, then we are still in the realm of seduction; *charme* begins when there are no longer any secrets. Hence there were Dionysian mysteries, whereas Venus never had them: "In her role as scarecrow," writes Radiguet, "Venus lacks authority"![37] All this leads one to believe that the notion of purity that underlies venusian *charme* (and perhaps all of Roman religion) is completely different from that implicit in Greek religion. In Greece *katharma* meant *pharmakos*, scapegoat, as well as purifying sacrifice. For Girard, this refers to a conception of purification as purgation, as the evacuation from the city of all that was held to be harmful by means of the exercise of a violence analogous to the violence from which one wished to liberate the society. *Pharmakon* implies an identity between the evil and its remedy.[38] In Rome, however, the substitution of *vinus-venenum* for blood seems to imply a concept of purity as a simulating operation, displacement and transfer free from passions and traumatic exclusions. *Venenum* could also be merely water tinted with red or myrtle wine, like that used by matrons for cleansing themselves in the Veneralia feast of the 1st of April, dedicated to Venus Verticordia! Whoever conforms to rituals and scrupulously carries out ceremonies is *castus*. The Roman ritual without myth dispenses with fixed contents having a precise identity. Purification seems to become precisely the contrary of purification in Greece: it is not the identification and expulsion of something held to be impure, but the ritual emptying out of *all* aspects of life. On the 1st of April Roman matrons celebrating the rite of Venus were as *castae* as the prostitutes who worshipped Fortuna Virilis.

We cannot conclude without mentioning the meaning of *venenum* that has prevailed in the history of the word: *venenum* as deadly drink. But here, too, it is difficult to avoid the impression that the Romans aimed at a displacement of death itself. Plutarch attributed to Numa Pompilius the institution of an ancient cult dedicated to Venus Libitina, goddess of funeral rites. He observes that the Romans presumably shrewdly assigned the regulation of the birth and death of men to a single goddess.[39] Such a cult appears to be inspired not by a tragic conception of existence, like that of the Greeks, but rather by an aspiration to make the cultural aspect of death coincide with that of birth. Nothing remains foreign to the venusian enchantment of rites and ceremonies.

The very etymological origin of *charme*, which comes from *carmen*, refers to this perspective. *Carmen* has the general meaning of a cadenced formula, endowed with formal characteristics artificially regulated and maintained independently of their original meaning. Both religious formulas and the text of the law were called *carmen*. In the ritualism of the *carmen*, Roman religion perhaps finds its own unity;[40] in the *charme* of the quotidian, the contemporary crisis perhaps finds its own solution.

TRANSLATED BY BARBARA SPACKMAN

MARIO PERNIOLA
Decorum and Ceremony

I. The Resplendent

What is the relation between beauty and effectiveness, form and action, aesthetics and politics in classical antiquity? The link seems implicit in two concepts: the Greco-Roman *prepon*, *decorum*, and the typically Roman *caerimonia*.

The history of the first concept is quite complex and tortuous. The word and originary meaning of *to prepon* come to ancient Greece from the vision of an efficacious beauty that appears distinctly before our eyes, that distinguishes itself by its perspicuity, that excels, shows itself, shines, imposes itself on the gaze and glows in its singular reality. The Homeric hero, for example, is endowed with this quality: his virtue is visible, it falls beneath the gaze of all, stands out conspicuously, distinguishes itself without concealment or dissimulation. He affirms himself independently of and prior to any distinction between appearance and substance, seeming and being.[1] It is noteworthy that among the Greek words that indicate the beautiful, only *to prepon* is etymologically connected to an Indo-European root whose fundamental meaning refers to appearance and vision:[2] not *to kalon*, in which the idea of beauty seems to be etymologically connected to health, to the correct proportions of the limbs; not *to agathon*, whose originary sense seems to be linked to force and power, and hence to courage and nobility; nor, finally, the noun *ho kosmos*, which means order. Not even *to agalma*, which indicates the ornament and then the statue of the gods as opposed to *hē eikōn*, the statue of men, seems closely connected to vision.[3]

The first meaning of the verb *prepein* is instead "to be resplendent." In it the experience of the beautiful is wedded to the festive visuality that characterizes ancient Greek religion, which has been defined as the clairvoyant knowledge of festive man in which seeing is no less important than being seen, in which the knowledge of the divine takes on the form of an epiphany, of a radiant manifestation of reality.[4] It is not by chance that Heidegger saw in appearance as splendor and brilliance, in the unconcealing permanent self-imposition of the phenomenon, of that which appears and shows itself in itself, the most originary and essential experience of Being in the West.[5]

The verb *prepein* refers to the inseparable unity between being and appearance, between that which is and that which shines forth, between beauty and effect. The poets use this verb in verses that firmly join together beauty, decision, and success. For Pindar, for example, "gold and righteousness are proved [*prepei*] for one who testeth on the touchstone."[6] At the center of Tydeus's shield, as described by Aeschylus, "the moon, queen of stars, the eye of night, shines [*prepei*] radiant" (*Seven against Thebes*, 390). Finally, the only time that Plato uses the verb *prepein* with the meaning "to be resplendent" is at the beginning of the *Republic*, where it refers to the beautiful procession with which the people of Piraeus celebrate the festival.

II. The Appropriate

The originary unity of being and seeming, of effectiveness and beauty, is nonetheless shattered by historical experience, which shows that what is resplendent and what effectively succeeds do not always coincide. *Prepein*, to be resplendent, in its originary autonomy is no longer sufficient to guarantee victory and historical success. Beauty that would maintain its relation to reality must "adjust to," "befit" that which is other with respect to it. This is precisely the second meaning of *prepein* which takes hold and is maintained in Greek and upon which is grafted the problematic of *to prepon*, understood as that particular type of beauty that adapts itself, that is appropriate. Thus, by virtue of its relation to the other that constitutes it, it is in opposition to the absolute and universal conception of the beautiful that is implicit in the canon.

The lyric poets once again elude the tragic experience of this split by saving the autonomous splendor of the beautiful for poetry: Sappho writes, "It is not right that a cry be heard in the house of those who serve the Muse: mourning does not befit [*prepoi*] us."[7] But from the moment when, as Thucydides says, "good counsel clearly expressed is open to suspicion no less than harmful counsel" (Thucydides, 3.43), the divorce between that which is resplendent and that which succeeds is wholly consummated. The resplendent finds itself in the midst of a battle in which it has no advantage, and in which, in fact, it is likely to give in. Only by adapting itself to circumstances better than its adversary, only by knowing better than the enemy what is and what is not *prepon*, what is appropriate and what is not, "what must be done in the right moment" and what must not be done, can it continue to be resplendent. The notion of *prepon* is thus wed to the more ancient notion of *kairos*, occasion (timeliness). Though this link was already implicit in Pythagoras, especially when he maintained the appropriateness of delivering childish speeches to children, womanly speeches to women, archonic speeches to archons, ephebic speeches to ephebes,[8] it is only in Gorgias that the connection between what is appropriate and occasion is freed of the originary mystical meaning of *kairos* as referring to the harmony of the cosmos. In Gorgias, the fundamental advantage accruing to the Pythagorean sage from his knowledge of the essence of being, whence his *polutropia logou*, his ability to express the same thing in many ways, is dissolved. The beautiful, taken in the Greek sense which implies the true and good as well, is thus forced to use the arms of its enemy: "Hence not only must he who would advocate the most dangerous politics enter into the good graces of the people, but he who counsels the better way must also secure trust by means of trickery" (Thucydides, 3.43).

For Gorgias, the problem of *to prepon* is essentially the problem of language and its powers of seduction (*apatē*):[9] "The word, like the proclamation sent forth at Olympus, invites whomever it wants, crowns whoever is capable."[10] But why must the resplendent have more *apatē*, more seductive force, and hence succeed better? Gorgias's answer is a drastic one: there is no *prepon*, no resplendent that is not appropriate, that does not conform to the occasion and have seductive force sufficient to assert itself and win. The resplendent thus seems to be completely crushed beneath the heel of the effective, to the point of complete identification of the two: all that is real is also beautiful because it conforms to the occasion and precisely by virtue of that conformity was able to become real. No matter how tragic Gorgias's position may appear because of the impossibility of accepting the identity of the beautiful and the real,[11] this identity is always mediated for him by *apatē*, by the seduction of the word. The resplendent is effective and the effective is resplendent only where there are people who are sensitive to the charm of the word, only where, as in Greece, there exists an experience like the great tragic tradition in which "he who succeeds in a deception better conforms to reality than he who fails, and he who is deceived is wiser than he who is not deceived."[12]

But is it still possible to identify the resplendent with the appropriate, the beautiful with the effective in a country where people are not used to listening to speeches attentively, or easily forget what they have heard, or, like the inhabitants of Thessalia, are too crude, too lacking in good sense (*amathesteroi*) to be deceived by the word, by this powerful sovereign who with "a tiny and completely invisible body accomplishes profoundly divine works"?[13] In Xenophon's Socrates the use of the adjective *prepōdēs* slides toward a definition much closer to the useful and appropriate than to the beautiful: "The most appropriate place for temples and altars is an open and completely isolated place" (Xen. *Mem.* 3.8). Xenophon's Socrates completely identifies the beauty of a building with its utility, with its being *khrēsimos*, or even better *harmostos* (from *harmozein*, which means to adapt). Xenophon's Socrates thus inaugurates a functionalist conception of beauty that is completely foreign to the original identity of beauty and effectiveness implicit in *to prepon*. *Harmostos* corresponds to the Latin *aptus*, and it is precisely as appropriateness to the goal that decorum is understood by an entire tradition of thought that is developed above all in the medieval period.

Even more radical is the negation of *prepon* worked by Plato. Plato inaugurates a complete separation of substance and appearance, of seeming and being. Against Hippias, who advocated the identity of the beautiful and the appropriate of *kalon* and *prepon*, the Platonic Socrates clearly separates the *kalon* from contingent effectiveness and proposes to search for "a beauty . . . that will never appear ugly to anyone anywhere," that is, "beauty itself, that which gives the property of being beautiful to everything to which it is added—to stone and wood, and man, and god, and every action and every branch of learning."[14] In the name of the *eidos* of beauty, Plato addresses a radical critique to the Sophists, whom he accuses of making objects seem more beautiful than they are in reality and hence of deceiving their listeners, and concludes, "It cannot be the appropriate, for on your own view this causes things to appear more beautiful than they are, and does not leave them to appear such as they are in reality."[15] The conception of beauty itself opposed to *prepon* is, in fact, only an aspect of the broader and more general reduction of Being to entity that inaugurates Western metaphysics, in which there is no place for a beauty that is also effective, for a historical resplendent that wins because it is resplendent. According to Plato, "The same cause never could make things both appear and be either beautiful or anything else."[16] Beauty is always beautiful whether it wins or not.

The metaphysical negation of the appropriate is reiterated by Plato in the *Ion*, where he speaks of poetry. To Ion, who boasts of knowing what is appropriate for man to say, for woman, servant, and freeman, for those who order and those who obey, the Platonic Socrates opposes the necessity of distinguishing the word that comes from true knowledge and the poetic word, which, by divine lot yet knowing nothing, can say so many beautiful things and do no wrong (*Ion* 542a).

A defense of the notion of *prepon*, or the appropriate, is found instead in Isocrates in a form that nevertheless distances it from the oratorical activity in the assembly or courts, and that gives it a new sense in relation to *kairos*, to occasion. A disciple of Gorgias, Isocrates repeats the idea that "speeches cannot be beautiful unless they are in accord with the circumstances, adequate to the subject and full of novelty" (*Soph.* 13). In fact, he chided Socrates and the Socratics as much as the masters of improvisation like Alcidamas for overlooking the mobility, variety, and diversity of the human situation by superimposing fundamental schematic forms: for the Socratics, ideas; for the orators, rhetorical techniques. Isocrates compares these schema to the letters of the alphabet. In this way, he attempts to separate the problematic of *prepon* from the risks that arose both from Gorgias's too-empirical formulation and from Plato's implacable criticisms. This is accomplished by means of two fundamental innovations: the connection of *prepon* to the problematic of *paideia*, education; and the adoption of a Pan-Hellenic point of view. Thus are born the humanistic interpretation of *prepon* and the constitution of the orator as subject. While for Georgias the orator persuades and wins the more he makes himself nothing and no one in order to adapt himself to varying occasions, for Isocrates the orator, having become not only master of oratory but also master of life, derives persuasion instead from being trustworthy, from the acquisition of a moral status that elevates him above politicians and judiciary writers.[17] It is not coincidental that Isocrates defines himself as philosopher and considers Socrates and the Socratics to be sophists. Pan-Hellenism, the unity of all Greeks against the barbarians regardless of local struggles between individual cities, confers a political content upon this solemnization and self-promotion of the orator and allows him to become the defender of an appropriate that is above quotidian effectuality but still bears a much closer relation to it than does the Platonic "beauty in itself."

With Aristotle the readjustment of the claims of rhetoric to a search for means that may be persuasive for every argument, and the determination of its object of study as the probable or that which appears probable, breaks *prepon's* link both with the beautiful, to which Aristotle attributes an autonomous dimension, and with the effective, since for Aristotle "things that are true and things that are just have a natural tendency to prevail over their opposites."[18] Nevertheless, given that the many are not capable of learning the principles of science through teaching and, because of their moral baseness, are persuaded by things external to pure and simple demonstration, one must take the factor of the appropriate into consideration in elocution.

The appropriate manifests itself as propriety, as adequation of elocution to the emotions, characters, and arguments with which it deals (*Rhetoric* 3.7.1408a10). Above all, the adequate representation of characters (*ēthē*) is important to the ends of the concept of *prepon*. In fact, it inaugurates a third way different both from that of the absolute indetermination of the *kairos* of Gorgias and Isocrates, and from the schematic abstractness of the rhetoricians; it is aimed at determining as many "appropriates" as there are categories of people identified concretely and historically on the basis of their *ēthē*, their customs. Aristotle writes:

> Furthermore, this way of proving your story by displaying these signs of its genuineness expresses your personal character. Each class of men, each type of disposition, will have its own appropriate way of letting the truth appear. Under "class" I include differences of age, as boy, man, or old man; of sex, as man or woman; of nationality, as Spartan or Thessalian. By "dispositions" I here mean only those dispositions which determine the character of a man's life, for it is not every disposition that does this. If, then, a speaker uses the very words which are in keeping with a particular disposition, he will reproduce the corresponding character. (*Rhetoric* 3.7.1408a25–30)

In this Aristotelian interpretation, the notion of *prepon* loses its originary meaning as effective resplendent because it is recuperated in the context of a problematic of representation. It is not by chance that it is given another important application implicit in the *Poetics*, where Aristotle speaks of the character and qualities of characters of tragedy (*Poetics* 1454a16). The notion of *prepon* also loses the aesthetico-political tension that characterized the positions of Gorgias and Isocrates. For Aristotle, the appropriate is accessory: "One may succeed in stating the required principles, but one's science will no longer be dialectic or rhetoric, but the science to which the principles thus discovered belong" (*Rhetoric* 1.1358a25). The Aristotelian appropriate can thus constitute the point of departure for that aesthetics of the characteristic that, through Theophrastus and Horace, will develop in opposition to the classicist aesthetics of the canon right down to romanticism.

The final attempt by Greek thought to think the beautiful and the effective together is that of Panaetius of Rhodes. In this thinker, however, the word *prepon* is eclipsed by *kathēkon*, derived from the Stoic tradition and referring to appropriate action as opposed to *katorthōma*, virtuous action in the absolute. While this last comes only from *logos* and hence is the exclusive patrimony of the wise man in the Early Stoa, *kathēkon*, understood by Zeno in its etymological sense as that which befalls, happens to, descends on someone, can be accomplished even by fools. Panaetius brings a fundamental correction to this tradition by emphasizing the importance of *kathēkon*, of the appropriate in a circumstance, in comparison to *katorthōma*, absolute duty. In addition, he interprets Zeno's precept of living according to nature in a very specific and personal sense and considers *kathēkon* precisely as action that is appropriate, that conforms to one's personal nature. The notion of the appropriate that in Zeno is something that supervenes and happens, is thus internalized, allowing Panaetius to attribute to it that beauty that the most rigorous Stoics assigned exclusively to virtue.

The relationship between beauty and effectiveness that constitutes the conceptual node of the notion of *prepon* seems nonetheless to be weighted in the direction of a social reality that owes its *raison d'être* to factors that have nothing to do with either virtue or beauty. Thus the beautiful appearance of the individual personality, the harmonious and elegant style of life,[19] appear to be something added to an effectiveness that has become such independently, rather than being the cause of effectiveness or an inseparable part of it. This is in keeping with Panaetius's assertion that happiness requires not only virtue (as Zeno and Chrysippus hold) but also health, wealth (*Khorēgia*), and strength (Diogenes Laertius, 7.128).

III. Decorum

Cicero is the great interpreter and expositor of Greek theories of *prepon* in the Latin world, and in particular of the oratorical version given it by Isocrates and the moral interpretation given it by Panaetius. Without entering into the vexed question of Cicero's originality in relation to Greek models, it is important to dwell upon the Latin word Cicero chooses to translate the Greek and upon the consequences of this choice. After some hesitation, that word is *decorum* in both *Orator* and *De officiis*.[20] Etymologically, *decorum* has nothing to do with *prepon*. While *prepon* originally refers to the unity of vision and effectiveness, the Latin *decorum* implies instead a connection between behavior and effectiveness. *Decorum*, in fact, comes from the impersonal verb *decet*, related to the Vedic "dȧsti" whose meaning is "to pay hommage to" and whose source is traceable to the Indo-European root * dekˆ- (to take, to accept, to salute, to honor). Though essentially different from the Greek, *decorum* too thus refers to a religious experience based not upon the festive visuality of the divine but rather on the acceptance as one's own of the will of the gods, of listening in order to grasp the signs of *fatum*, on repetition and veneration. In fact, in Latin the idea of beauty is associated with religious rites much more than with vision: *pulcher* has a specifically religious value in the language of auguries and designates any favorable omen gathered from the observation of birds or the examination of entrails. It is also applied to favorable divine powers, to beings favored by the gods, and to the effect of divine will and evokes in all cases a prosperity attributed to the gods.[21] The case of *venustus* is analogous. Hence there is in Latin as in Greek an originary inseparability of the beautiful and the effective. But whereas the Greek connection between that which appears and that which is effective cannot be maintained and, by clashing with historical experience, produces a metaphysical solution that completely separates appearance from reality, the Roman solution is radically different. The identification between beauty and ritual behavior renders even firmer the Roman link between the appropriate and the effective.

This is apparent above all in the way Cicero develops the notion of *decorum* in *Orator*. The problem of what is appropriate and what is not (*quid deceat et quid dedeceat*) is completely divorced from any external evaluation and considered in itself. The important thing is to convince, to entertain, to move: it is up to the discernment of the orator to know how to judge what is expedient in each case and how each case should be conducted. Success is in no way separable from that which is fitting (*decet*): "For the same style and the same thoughts must not be used in portraying every condition in life, or every rank, position or age, and in fact a similar distinction must be made in respect of place, time and audience. The universal rule, in oratory as in life, is to consider propriety" (*Orator* 21).[22] *Decorum* is determined by three elements: by the "re de qua agitur," by those who speak, and by those who listen. The orator must therefore be master of all three types of oratory: Attic, which is plain, without ornament and characterized by a *negligentia diligens*; the middle style, rich in figures and mutations yet serene and tranquil; and the grave and adorned style, opulent and magnificent. Whoever begins with too much ardor before an unprepared audience will undoubtedly offend *decorum*; such behavior will have results contrary to those desired "Furere apud sanos et quasi inter sobrios bacchari vinulentus videtur"

("He seems to be a raving madman among the sane, like a drunken reveller in the midst of sober men") (*Orator* 23). Cicero states that he admires above all those who know what is fitting to the case: the essential quality is to know how to adapt words to people and times, for one must not speak always, nor in the presence of, nor for everyone in the same manner.

A fundamental portion of *De officiis* is devoted to the concept of *decorum* and merits a specific and detailed study. In such a study, it would be important to identify at least four elements for reflection. In the first place, Cicero's difficulty in distinguishing between *decorum* and the *honestum* is evident: such a distinction "facilius intellegi quam explanari potest" (*De officiis* 1.27.93). Indeed, we can trace back to Cicero the formulation of the notion of "je ne sais quoi," of the "nescio quid" that is so important in the modern development of aesthetics. Second, it is important to point out the semantic and conceptual slippage worked by Cicero in his translation of the Greek *kathēkon* as *officium*, which comes from *opus* and is in turn closely connected to religious rites. Third, this shows that the Romans' attitude toward historical reality is identical to their attitude toward the divine. The fundamental intuition that gives rise to *decorum*, *quod decet*, never enters into contradiction with historical experience, as happens instead in the case of Greek *prepon*. In the fourth place, it is perhaps possible to identify an influence of the Ciceronian notion of *decorum* in the history of Western culture, an influence that precisely because it is subterranean is perhaps more determinant and effective than the Platonic-Aristotelian metaphysical tradition that is held to be the royal road of Western thought.

IV. *Ceremony*

The link between form and effectiveness, appearance and ritual implicit in the Latin concept of *decorum* is even closer in the typically Roman notion of *caerimonia*. In order to understand this notion, we must first free ourselves of the spiritualistic prejudice that considers ceremony as stereotyped, superfluous, residual, idolatrous, maniacal, desperate behavior, seeing it as formalism and sclerosis, lacking in depth and substance. This prejudice is activated every time *caerimonia* is thought of as mere *carimonia* (from *careo* = to lack, to deprive oneself), according to an erroneous etymology already formulated in antiquity.

On the basis of several passages from classical authors, Karl-Heinz Roloff, the author of the most extensive study of the Latin word and concept *caerimonia*, shows that in addition to meaning action and ritual behavior, the term also designates the very being of the divine, the object of religion, or what we can roughly translate as "holiness" (*Heiligkeit*).[23] That ceremony means much more than sanctity seems proved by a passage from Suetonius that juxtaposes the *sanctitas* of kings to the *caerimonia* of the gods, thereby pointing to the difference between their modes of being. Hence the word does not refer to a lack but, on the contrary, to the fullness of the sacred. This explains why the word is always used in the singular with this meaning, so much so that late grammarians took it as a *plurale tantum*.

When Cicero speaks of a *caerimonia legationis*, and Tacitus speaks of *caerimonia loci*, they are thinking above all of the being of the thing itself. Finally, Caesar, in *De bello gallico* (7.2), when recounting the conspiracy plotted by the Carnutes together with other peoples of Gaul, says that they refused to exchange hostages in order that their plans not be revealed, but requested that, having united their respective military insignia in a single fasces, they all pledge by word and oath not to split off from the others once the war had begun. Caesar defines this act as a *gravissima caerimonia* because the military insignia united together acquire a sacrosanct, objective power, independent of the beliefs of man and such as to effectively take the place of the exchange of hostages.

If, then, one must speak of the *ceremoniality* of the sacred, it cannot be understood as *festivity* (*Feierlichkeit*), the festive and festal contemplation that Károly Kerényi attributed to Greek religion, considered in its connection with vision, manifestation, the splendid appearance of the phenomenon.[24] There is no reference at all in Caesar's text to an epiphany of the divine. The Carnutes' attention is completely concentrated upon the action that they are about to undertake and the necessity of founding such a historical action, full of risks and unknown factors, upon obedience to the *gravissima caerimonia* of the insignia joined together. Such a ceremony is not a festivity in honor of an alliance but the objective, extremely serious, and binding guarantee of an alliance.

Even less can "ceremoniality" be understood as "spectacularity." *Ludi scaenici* are foreign to archaic Roman religion, to which the word used by Caesar refers.[25] A great indeterminacy with regard to the gods is characteristic of Roman religion; often one does not even know if they are male or female, and their identities are reduced to their names. Dumézil has rightly compared the Roman pantheon to a world of almost immobile shadows, to a twilight crowd in which it is difficult to distinguish clear forms.[26] Even though Caesar was speaking of the Gauls, it is clear that he attributed to them a typically Roman way of thinking.

In its most profound meaning, *caerimonia deorum* does not refer to the cult that belongs to the gods, of which the gods are master, nor to the cult dedicated to them, but rather to the *externality* of the mode of being of the sacred. Here a meditation on the Romans encounters the theory of the sacred as "*completely other*," as *difference*, as a radical refusal of any anthropomorphic conception of the divine. Such a convergence of a theory of the sacred that has roots in the most radical monotheism with the Roman paganism that Hegel considered one of the most deplorable forms of superstition is indeed surprising. And yet the objective convergence, despite the distance that separates Roman ceremoniality from Yahweh, of Jewish iconoclasm and the noniconism of early Roman religion, which, according to tradition, knew no sacred images for the first 170 years of its

history, is perplexing. It is, in any case, essential to understand that ceremonial externality is precisely the contrary of a panoramic and decorative
mode of being.

The subjective meaning of ceremony, understood as ritual operation
and behavior, is equally as important as the objective meaning attributed
to "things themselves." The former sense of the term, which is the most
widespread and common, is nevertheless strictly connected to the latter.
In the ceremony recounted by Caesar, for example, there would have been
no objective holiness if the act of uniting together the military insignia,
an act directed toward a determined goal, had not been carried out: "The
action itself is, in this respect, holiness; without it there would be nothing
sacred, but on the other hand only there where the *signa* are together is
there the sacred."[27] Thus not only is *externality* the fundamental characteristic of the divine being, it is also and by the same token the essential
character of the religious rite, which has no need at all to ground its validity upon a belief, a myth, or internal experience. Here the distance between Roman religion and the theology of difference of Judaic origins
appears clearly: in the former the exteriority of the rite corresponds to
the exteriority of the sacred, while in the latter the exteriority of God corresponds to the interiority of the cult.[28] This does not mean, as Hegel
says, that Roman ceremony breaks the individuality of all spirits, suffo-
cates all vitality, and is linked to a total emotional and spiritual insen-
sitivity. The relationship between exterior and interior is overturned: it is
not interiority that grounds and justifies the cult, but rather ceremony,
the extremely precise and scrupulous repetition of ritual acts that opens
the way to a nonsentimental, nonintimistic sensitivity that is not thereby
less articulated and complex. In Caesar's story, ceremony creates a firmer
solidarity among the conspirators than would have been guaranteed by
exchanging hostages. Such a solidarity is not exclusively religious, but at
the same time political and juridical.

One cannot completely grasp the Roman meaning of *caerimonia* if one
overlooks the politico-juridical dimension, which, however, is not to be
understood as *lex*, as an act that is voluntarily binding, but rather as *ius*,
as a rigorously technical rite, a procedure in which both the magistrate
and the parties involved play already rigorously determined roles. The oblig-
atoriness of the ceremony depends not upon the subjective consensus
of the participants but upon the magistrate's ability to unite the particular case to the general and abstract form of the rite. *Ius* is an *ars* that "in
sola prudentium interpretatione consistit."

Ceremonial behavior is thus determined in relation to two external
and objective terms: the particular situation and the ritual form. *Prudentia*
is the capacity to harmonize them. On the one hand, obedience to the
occasion, to the particular, to opportunity, does not dissolve into mere
opportunism because it is accomplished with reference to a frame, a general scheme inherited from the past. On the other hand, obedience to
ritual is not mere sclerosis because it aims toward the solution of a ques-

tion, a concrete problem. This harmony between form and occasion is a recurrent theme in Jhering's major study of Roman law, which he defined as "the system of disciplined egoism."[29] The practical instincts of the Romans, Jhering claimed, had made rules and institutions so elastic that even when scrupulously observed, they always adapted themselves to the needs of the moment.

The concept of externality as referring to the Roman world does not mean transcendence of a law that imposes itself unconditionally upon human interiority. Ceremony is not the execution of an eternal and immutable "ought-to-be," nor is it the actualization of a metaphysical mystery; the terms upon which it grounds itself are all objective but immanent to history. The Roman sacred has no pantheistic or mystical character: as Roloff has observed, it exists above all in each individual case, in each individual event, fully conforming to the "casuistic" attitude of the Roman way of thinking.[30]

Ceremony is the opposite of decoration, spectacle, *mise-en-scène*: it reveals itself as the condition of effectivity, operativity, history. This is particularly clear in the Roman conception of time, which is as different from the eternal return of primitive societies with their cycles of ritual death and rebirth as it is from the linear history of Judaism with its messianic tension toward final redemption. In Rome, the *ceremoniality* of time is embodied in the *calendar*, a formal structure of days, months, and holidays that always returns without hindering the historical activity of men but rather furnishing an indispensable point of reference for the chronological identification of every action both in deed and in memory.

The cyclical time of primitive societies, in which what counts is the re-actualization of the original mythical archetype, and the linear time of Judaism, which considers the deeds of Israel to be the deeds of God himself, are both *mythological* times, times in which there is an inseparable link between the chronological dimension and its content. It is precisely this connection that grounds the holiness of these experiences of time. The Roman calendar, instead, grounds a demythified but not therefore desacralized or insignificant time. It furnishes a frame, a network of reference points, a texture whose elements are sacred but which does not say a priori what they must contain, nor transform their contents into a sacred history a posteriori. The ceremonial structure of the Roman calendar presents itself as a condition of history: first it leaves undertermined the concrete nature of the event; then, when the event has occurred, it does not annul its specificity by inserting it into a process whose ultimate meaning is final redemption but goes about maintaining it by making it a "precedent."

The ceremoniality of time is, finally, a transfer from the same to the same. There is nothing to teach or to learn but procedures, ceremonies, rotating movements in which the occasion, the most empirical particularity, the specific situation, must be played out. It is useless to try to escape from "Mamurius's game": one must continue to play despite the blows.

The blacksmith Mamurius's teaching opposes that of the other Indo-European "lords of fire": not *wut*, religious furor, wrath that terrorizes enemies, but rather calm, indifference, mimeticism, in a word, ceremony.

TRANSLATED BY BARBARA SPACKMAN

PIER ALDO ROVATTI
Maintaining the Distance

I would like to thematize two of the possible meanings of *scarto*. One of the definitions of *scarto* indicates an abrupt dislocation or swerve, a clear-cut passage, like a change of scene. There is a spatial element in it, both in the strict and in the figurative sense: it is a passage from one place to another. We could say that in the uniform or the apparently uniform movement of experience something occurs to modify the order, and we need this disordering event in the same way that we need to change place when we have stayed somewhere at length and monotonously so. But a second meaning of *scarto* is distance: the spatial reference remains but the accent changes. Distance: we can think of the gauge of a railway track. A "distance," a gauge, which can be reduced but not filled. An interval, therefore, like a zone of separation, an inability to meet: a place interposed between us and ourselves. Also a void, or simply a *not* which can become laceration, cleaving, a hole, or an abyss. We might ask ourselves, do we also need the lack that this second sense of the term brings to light?

In order to verify the first statement and attempt to answer the above question we must broaden our perspective. If we look at the main components of today's philosophic thought we can say that our philosophy has always oscillated between affirmation and negation, between affirmative thought and negative thought. In affirmative thought everything—the whole of reality—is a plane of consistence which is homogeneous and, at first glance, without flaws, though also a flowing plane upon which events slide. Affirmative thought, moreover, adds the idea of power to consistence and flowing: reality is something that develops, that contains a dynamism, a creative activity within itself. The real, then, is something complex that dilates in forging ahead. But not according to the most obvious and simplest image of a roll that unrolls, but one of ramification, bifurcation, differentiation. Power is the power of ramification; the impetus forward produces differences. The passing of things is only apparently a tranquil flowing. The image or metaphor of the river is apt: the river that seems to flow evenly downstream does not simply follow its inclination but is troubled incessantly by disturbances and fluctuations.

The course of the river is broken by turns, by unexpected bends; its waters rear and stop in midstream in order to avoid obstacles. We call these disturbances obstacles of the course and by now we have learned to recognize them; we know, that is, that they are not marginal elements: they coincide with the course itself. It is illusory to think of a flowing without fluctuation. On the contrary, it is precisely this complex of fluctuations that we improperly call flowing. The bends of the river are its going ahead. If coming downstream recalls physical entropy, then going upstream—which paradoxically is the bends, the turns of the river—is the negative entropy that is life. What is the negative within this perspective? The negative is here indicated by the image of inclining; it remains as an almost imperceptible backdrop, a tendency—certainly—which, however, countertendencies mask and impede, almost hiding it. There are no falls. There is only flowing. The pure negative would here be flowing in its pure state, simply an appearance. It would be the tendency toward order, but the river does not follow a direct route. Its tortuous course is also experience, which is almost always the longest path, made up of curves and bends.

If we turn now to negative thought, the image which we can refer to is that of a narrow catwalk, of a fine wire that a tightrope walker—that is to say, ourselves—must cross. A narrow passageway between the banks, although in this case the passage from one bank to another is not the essential thing. What counts, instead, is what takes place between the banks: the difficult equilibrium, the risk of this crossing, which is man's experience. If the plane of consistence presents itself as a *plenum*, the *plenum* now becomes a lateral element, secondary and fragile; a minimal point of support that can be reduced to the wire on which we seek equilibrium. The void, on the other hand, is endless, pervasive. The abyss is no longer only beneath us but is everywhere. Man's experience is the risk of sinking, of being engulfed, of vertigo. It can produce blockage, paralysis, the feeling of nonpower—that is, of impotence. It can produce the contemplation of the abyss, or else a provisional morality and logic. It can produce the expectation of the worst or of the next-to-worst; the speculation of risk or the reckoning of risks. Within this perspective, man is fundamentally lack: he misses himself, he misses reality as if he has missed a target. He can only, and provisionally, deceive himself. Deep down man knows that this equilibrium is extremely precarious: a small movement, a *scarto*, is sufficient to lose it.

Perhaps, however, neither of these positions is tenable as such, neither is "the position." We cannot convince ourselves that there are no discontinuities as the idea of the plane of consistence and flowing would lead us to believe, making us think that everything is a question of weft—thicker, wider, direct, oblique. But neither can we think that experience is really only the knowledge of the void, that it is obsessively dominated by this lack, that it is only a way of facing it. Contemporary thought probably oscillates between these two beliefs, and we are continually tossed from one to the other.

But let us return to the two images of *scarto*. The dislocation, the change of scene, is a sudden happening; something suddenly takes place like wakening from sleep, like a new fact that changes and gives a different tone to our experience. We might ask ourselves: can we plan something that suddenly happens? In other words, can we find a position from which to observe without interfering, without straightening out the curves? Is there a way of looking at the river as fluctuation? Or is our "looking at" always destined to transform the fluctuation into a flowing? If we do not wish to reply with a downright no, it is then the course of consciousness that we must set about correcting, "dreaming" a consciousness that does not imprint itself upon experience in order to render it linear, a reason that in its ancient rigor has recognized rigidity and, therefore, in its ancient strength, or in that which it presumed to be its strength, an inefficacy. A reason that in some way starts to circle around its object, that constructs a less drastic knowledge, and looks for a mobility of its own and, at the same time, a syntony with its object. The dislocations, the swervings are precisely that object.

A rationality that must now force itself to acknowledge a rhythm that is not a logic; that will probably have to make itself smaller given that the *scarto* at which one looks and with which one must enter into syntony—is minimal. It is not, in fact, a "great" change. A reason that will not have to render itself completely manifest because the *scarto* as such, in its sudden happening, eliminates this possibility. A reason that will have to cease telling itself: "I am," because the Being of the *scarto* is fleeting and evasive—it is not there, neither is it before us, it is not an entity. It is not "this" or "that"; it has no precise border or limit. Hence consciousness will have to learn to live without this precision. And it will also have to get out of the habit of being before the object, in the foreground, of being unique and equal. In short, it will have to recognize itself in these *scarti* and, in order to do so, must substitute the "once" with the "repeatedly," definition with narration.

Is this possible? In other words—and formulating the question is very important—can an identity always be different, continually change place? We can hypothesize that for this to happen there has to be a doubling, a split; consciousness must double itself. It will have to render its claim image, imaginary. It must limit its automatic polarizing toward the one, establish itself in a play of forces, and yet at the same time stay to one side—not, however, on the side it has already learned to acknowledge as its own. The "one" does not disappear, the element of "once" still remains: the place changes within a possible topological description of consciousness. The paradox is precisely that the place where we retain our own *proprium* is not to be found.

What, after all, is this consciousness? How does it appear? It is not utopian, lucid, and transparent but neither is it a pessimistically opaque consciousness. It is not just identical to itself, or even merely an escape into difference. It is here that an extremely arduous task begins: a more complicated subject can be constructed, not only in order to account for experience made of movements, of dislocations of *scarti*, but in order to have experience of one's being reasonable, to be able to give a "status" to reasonableness in this situation.

At this point, however, let us look at the second form of *scarto*: *scarto* as the distancing from itself. It is inevitably a form of loss, a missing of the target. Negative thought tells us there is a lack. Weak reason, correcting, interpreting, and deciphering, might say there is a distance to maintain. Negative thought warns us that the *scarto* is not only a change of direction but also always a sinking. Can we believe that the ground does not give way and that everything happens in a place, in the place, that there are merely transfers to do, trajectories to change, or also (only) leaps to be made, spaces to reduce, and that it is always and only a matter of covering a route, a path?

Even if we move very slowly we believe we are advancing, moving forward. At the same time, however, we have evidence of something that holds us back, of a time, no longer a place or a space, that never manages to pass—of a leap that is never sufficient, no matter how small it is. A *scarto* that is never enough, no matter how literal it may be. And in any case, something else has already taken place on another level, downward perhaps, or backward. The void has crept in, has ballasted the thrust and has trammeled it.

But at this point the question could radically overturn the sense of the description. Is it not precisely this void that has permitted the thrust, that made it possible for the thrust to proceed? A vertical *scarto*, let us say, has cut the horizontal one: consciousness has also given way, every move is *also* accompanied by a fall into the abyss.

In this philosophical context, we might compare the positions of Paul Ricoeur and Jacques Lacan on metaphor. *The Rule of Metaphor* by Ricoeur is a battle against the translation of the metaphor into something insignificant, into a simple ornament, although above all against the metaphysical element that Derrida sees as being reproduced necessarily in metaphor itself. Ricoeur maintains that metaphor is important both in order to comprehend the novelty, the incessant invention, to which language can give rise, as well as to be able to understand the relationship between this novelty and its truth value, between language and reality. There is no authentic element that can be substituted and reinstated, nor a *proprium* that may be seen as the unfigurable element that remains at the bottom of metaphor: there is, however, a possible and infinite play in which the similar and the dissimilar can be combined to give rise to novelties of meaning.

Ricoeur recalls that old Majorcan tales and fables always began with an expression which when translated rings as follows: "This—which I am about to tell you—was and was not." Ricoeur therefore introduces the "not," but he introduces it as an operator of the game of impertinence and dissimilitude on the plane of linguistic creation. An opening is claimed against the closure of a sole philosophical scheme which sees metaphor as merely the reemerging of the aporia of Platonic and Neo-platonic metaphysics: an opening out toward all the figures that, in point of fact, play upon this relationship of the similar and the dissimilar. Ricoeur says, however, that there also exists a return. Finally, metaphor in its signification as an essential instrument of knowing, as the impor-tant instrument of knowledge (and metaphor is, in fact, a displacement, a transfer), allows the possibility of being able to bring closer that which is distant: the distant concedes, resisting. "It resists": hence the not. The dissimilar allows itself to be approached, though in a tension, within a horizon that Ricoeur, in fact, calls tensional. This closeness is a coming closer once again; it is a return. But of what? The explanation furnished by Ricoeur is not convincing. It would be a phenomenological return of the precategorical world, of that level of reality, in other words, which comes before every code. But if this is the case, what, then, has happened with regard to the not?

We realize that we do not manage to resolve this "not" on the plane of the dissemination of sense or, to stay with the language of theories of metaphor, of interaction. Thus to Ricoeur we can oppose Lacan's the-ories, to the extent that the latter, however, is not specifically a scholar of metaphor. For Lacan, metaphor and metonymy act in a homologous way: the metonymic displacement and the metaphoric sinking are two faces of the same procedure, without return. The metonymic chain of sig-nifiers, "this and then this," functions precisely because at each link of the chain something—which we can call an irretrievable meaning—has fallen. Lacan and psychoanalysis, then, indicate an interpretation of the fallen meaning.

I wonder whether it really matters to recognize the psychoanalytical "nature" of this aspect of the theory: in my opinion it is important, above all, to note that the "not," in a position like Lacan's, implicates precisely that movement of sinking, conjugated, chained with that of the superficial "horizontal" slippage. In Ricoeur the "not," in a coupled relationship with the "yes" of identity, identical and at the same time different, on this "horizontal" plane produces meaning, metaphor as invention of new meaning—that of the live metaphor, in short. For Lacan it is precisely the loss, the sinking, that permits the production of meaning, of a meaning that is then registered on the chain of metonymic displacements.

These observations on Ricoeur and Lacan perhaps allow us to sketch the outlines of the problem. Both speak of *scarto, écart*. For Ricoeur, infinite variations are possible within what he calls a "stereoscopic" vision, a vi-sion that remains unexplored, however. For Lacan, however, lack is productive precisely because it is the symptom of unattainability, the negative permits "meaning." On the one hand, the impertinence of poetry, the nonpertinence of the poetic metaphor; on the other, the impossibility of pertinence.

But perhaps we cannot but see a nothingness acting in this "not": an art of seeing (such as the Aristotelian one recalled by Ricoeur) in which the subject, whatever it might be, involves a Lacanian being seen on the part of the object, and this being seen, an element of alterity in the vision itself, is the secret that makes the vision function.

Somewhere, then, a distance is to be maintained, not reduced. A "nothingness" is to be accounted for if we do not want this "nothingness" to reappear later and shake up everything. Coming nearer demands— rather, produces—distance, and this production required us to remove something; it demands a deduction, a tainting, a dislocating. The dislocations of experience also mean maintaining the distance.

TRANSLATED BY HOWARD RODGER MACLEAN

PIER ALDO ROVATTI
The Black Light

The Cartesian Gesture

The "black light" is a metaphor one encounters in some of Jacques Derrida's most important essays, in particular in his essay on Foucault's interpretation of the Cartesian *cogito* and in the essay dedicated to Lévinas.[1] In both it appears at the point of maximum theoretical tension and seems to mark a limit to philosophical reflection. Derrida's rereading of the *cogito* follows a movement, in both tone and theoretical approach, that is the opposite of calm acceptance of certainty. If natural light—the natural lumen of Descartes—accompanies and lights the path of knowledge, then the subject's interrogation of itself, precisely because it is a question whose answer is neither preestablished nor capable of being preestablished, from the very beginning must take leave of this guide and venture into a place that Derrida calls "an unheard-of excess." Here the philosopher, in search of a thread that might support the meaning of his own identity, can either deviate or lose himself. To deviate—in this interpretation—may mean turning one's back on the task; returning "precipitately" to the naturally luminous zone, betraying the question and pretending not to have "seen." But what was seen? Something that blinds sight and at the same time seems to be the enigmatic source of sight. Descartes carried out this precipitous return; without it, he would not have been able to join the *cogito* to science, nor would he have played that central role for modern thought that, without exception, the history of philosophy ascribes to him.

To lose oneself may instead mean giving in to the seduction of the shadow beyond the light: to let oneself be sucked into the negative world of night, of abyssal darkness, of the void. Or else to believe positively that another face of day exists: a retro-world by now disenchanted in another scene. The *cogito* also inaugurates this modern descent: Nietzsche can teach us to see it as the attempt to end the illusory and nihilistic domination of every higher world and of every *askēsis*. But Descartes does not lose himself; he "repatriates." Contemporary philosophic and literary thought will later take on the task of a "lucid" exploration of the night, forging ahead (for example with Freud) into the territory of the unconscious. And so this losing of oneself might simply mean duplicating the light as its opposite, maintaining the distinction between light and shadow by rendering them specular.

In this knowledge of the "negative," Derrida suspects an astute revenge of natural light capable of expanding and pushing its limits ever farther ahead, even if that threshold, once identified and not lost, necessarily poses a suspicion at the basis of every exclusive belief in knowing as light.

The metaphor of the "black light," then, is not so simple: the movement it would like to indicate differentiates itself equally from a step backward with the aim of conserving an acquisition of knowledge, and from a leap forward, to all appearances transgressive although in reality very often still guided by an "Enlightenment" faith in reason, held to be capable of enriching and modulating its luminous ray. As Derrida says, we have to try "not to extinguish that *other* light, a black and hardly natural light" to which Descartes paid his debt.[2] But how can we settle our debt with Descartes?

We can do it by acknowledging that, though he drew away from it immediately, Descartes touched upon the "hostile origin" of philosophy. The philosophical act cannot but be Cartesian in both its senses: in the authentic philosophical gesture, the subject can only move backward until identifying itself with the "black light," and the philosopher can only repeat—albeit differently—the Cartesian arrest and "return home." For the philosopher, identity cannot but be the loss of identity and, at the same time, the retrieval of identity. But does "identity" here mean the same thing in both cases? And does not the identity that unites them perhaps have a third meaning? What "rigor," what precision and communicability can be claimed by a consideration whose terms are so uncertain and oscillating, and which rotates around an obscure metaphor?

None, to be sure, if we continue to look for an objective definition of identity which places it at a distance, fixed, or in any case, endowed with a mobility contained within a stable frame, available to our cognitive sight in a way that we consider proper for an object of knowledge. The human sciences have elaborated many such definitions: why, then, turn our interest toward the uncertain hypothesis of philosophy? Simply because the question concerning identity is the question that is reproposed precisely in the attempt to weaken or eliminate the distance that separates us from an "object." Identity, as "representation," appears to us as the disguising of identity itself. Philosophy helps us to discover not only that between identity and representation there is an essential *décalage* but, and above all, that the interrogation concerning identity involves the meaning of all questions, even scientific ones: in fact, it implicates the very idea of meaning and, immediately, also that of knowledge.

Derrida's metaphor can perhaps serve as an outline. Knotted together in it we find many issues that can take us back to the primal scene of our philosophic project, that is, to the "heliotropic" genesis of philosophizing. But these same problems are produced in all their theoretical urgency in the present, post-Cartesian scene, already furrowed by numerous deviations and strayings: the ambiguous theme of this scene is constituted by the subject in the grips of the loss itself as center. The effect of truth that philosophy can propose for itself in this condition is undermined by the hermeneutic circle in which the very notion of truth in philosophy appears essentially to be entangled: nevertheless, this can be taken as a useful description of the theme. A description, not a definition; and utility in the sense of an approximation capable of producing a recognition and thus a horizon of communication. An age-old question, it appears to us to be so new as to take us almost completely by surprise— and to demand a theoretical invention regarding its every aspect.

Derrida sets into motion a triangulation between subject, knowledge, and light and warns us that the result cannot be contained within a concept. He points to language, moreover, as the place in which we can experience that coexistence of excess and the "return home" (the *rimpatrio*). Metaphor is a figure of language. Nevertheless, Derrida also warns us that the problem—as indeed Nietzsche has previously intuited—is an even more difficult one, that of the persistence of the metaphor in the age of its extinction: in other words, of something that we should perhaps no longer call metaphor. First of all, we should note the inadequacy of a structural description: not even the topological attempt (and here we can think of Lacan, who inherited the same problem from Freud and Heidegger), in trying to introduce a dynamic into structural fixity, seems to manage to come close. Models of this type succeed in accounting for the complexity of levels and the multiple play of effects. However, the movement that Derrida believes he reads—however stifled—in the Cartesian gesture does not seem reducible to a language of levels invaginated like the sides of a Möbius strip. The different modes of identity that coexist and oppose each other in the Cartesian gesture—to the point of making the notion of identity itself seem completely insubstantial—trace a movement of transformation that any model appears to betray totally. These various and contradictory identities would, in fact, cease to be so: the reason this does not happen is that they maintain themselves (or preserve themselves) as declensions of the "I am it"—that is, in that subjective displacement which is the very secret of the Cartesian gesture and which even Derrida takes for granted without realizing it. The condition of this gesture is to live it and to carry it out in person: all of the problems of description, of "violence" (as Derrida says in these essays), arise in an "unheard-of" way starting from the "unheard-of" character of a conversion which, on the contrary, seems so obvious as to be overlooked.

The "black light" is the linguistic mode with which Derrida chooses to indicate a subjective experience. The metaphorical enigma attempts to express the correlative enigma from a point of view that places itself on the side of the subject and that tries to assert its "disposition" as a decisive gesture: the philosophical act which, as such, has to be exposed to the greatest of risks (silence), and only to the extent to which it has to appear will the movement of identity and sense be given to be understood (a "view," which will not exactly be such).

A Precious Stone

Derrida himself exhibits certain difficulties regarding this question. If in the essay dealing with Foucault and Descartes the "black light" is to be safeguarded against the cancellations of a thought of simple transparency, in the text on Lévinas the theme of "excess," in relation to the metaphors of light, seems to be referred back to the normality of the metaphysical procedure. It has often been noted that "the heart of light is black."[3] Derrida is referring to the invincible connection between metaphysics and metaphor, captured in its inaugural moment: the Platonic metaphor of the sun. Here the light is already doubled: the metaphorical sun (of knowledge) is both the sensible sun and the ultrasensible, invisible sun.

The sun that we see with our senses rises and sets, shows itself and disappears, is now present and now absent; and it also allows itself to be perceived by men as a source of light and life—but only its effects, its luminous rays, are given to our sight, and not the source from which sight is blinded and which literally is not visible. Absence and invisibility set into motion the metaphoric displacement, the duplication of the suns, the infinite reference of the metaphysical, the invisible behind the visible, absence behind presence, metaphysics behind physics. The sun and metaphor unite in a chain whose links are the innumerable metaphysical variations and whose bond is "reference" itself, the beyond, the movement of transcendence, true reality which is concealed beyond phenomenal reality, the truth to which appearance refers. The Platonic sun is the true and the good: but its heart is black because it is not visible by the eye, which nonetheless entirely receives its "virtue" that is precisely the ability to see.

Metaphor, which with Plato assumes its philosophical status, certainly has a more ancient history in the horizon of thought and religious mysticism. An example that has recently attracted attention is the "black stone" belonging to the ancient Arabian gnosis. Studying the paradigm of the temple at the origins of Islamic culture, H. Corbin explains that the black stone corresponds to one of the corner columns of the Kaaba in Mecca.[4] These corner stones embody a complex symbology and fundamentally correspond to four types of light, each of which represents a prophet (Abraham, Moses, Jesus, and Mohammed). The Iraqi corner of the stone, the corner of the black light, is that of Mohammed, and within it, according to the Islamic mysticism of the Sabaeans of Harran, is concealed the very secret of the temple itself. Symbolized in it is the relationship between the terrestrial and the higher, intellective world. The symbol can be dissolved by way of a mythical account that takes us back to the first man. The black stone—as the Imam recounts to one of his disciples[5]—was originally the first of the angels: it is he who presents himself before God to vouch for the pact between God and Adam. However, with Adam's betrayal and his expulsion from Paradise, God gave the angel the appearance of a white pearl, which he threw down onto the Earth. Adam found the glowing pearl: the angel then announced himself to Adam and reminded him of the pact with God. At this point, however, God transformed the gleaming pearl into a heavy stone: Adam had to hoist it onto his shoulders and, bearing its weight, carry out the long journey from India to Arabia. The story teaches that in the terrestrial world the pearl can assume only dark and burdensome characteristics: it will be black but will have the power of arousing the memory of the angel and of God in the mind of man. Beneath the dark veil of worldliness man will be able to discover the luminous trace of divine intelligence. But let us remark that the stone is not worthless or vile but precious. It sparkles with red flashes, it emanates light; it is one of the lights—indeed the most important light—that supports the universe. And yet its heart is black. In this opacity man can "recollect," remember another light, unheard of, which is not human (natural) and which consequently will never be "visible."

Derrida's theme of the two suns, inferior and superior, seems to flow in the archaeology and the history of our thought: the black light, assumed within this paradigm in which metaphor and metaphysics continually take turns, corresponds to the absence, the fading, the eclipse of natural light—to its necessary referring to another light behind the light. As in the mythical story just mentioned, the black light is an intermediate element between world and ideality, the moment in which truth, in order to be revealed, has to deny itself sight, has to hide itself. It is an internal and essential moment for the metaphysics of light, which—according to Derrida—is nothing other than metaphysics itself. In this "dialectic" the mad audacity of the philosophic gesture disappears, that risky excess that knows how to tolerate the vigil of the powers of madness around the cogito.

When some years later (see *Writing and Difference*) Derrida tried to draw up a genetic-ideal study of the movement of the philosophical metaphor, the black heart of the light proved to be swallowed up in a "white" mythology, the place of the other metaphysical scene that the "flower" of philosophy—the heliotrope or sunflower—continuously recalls as its own life (whence it receives nourishment) but also its own death (where it is enchained): a play of mirrors, the one placed in front of the other, a "mise en abîme." As Derrida writes, "Metaphysics has erased within itself the fabulous scene that has produced it, the scene that nevertheless remains active and stirring, inscribed in white ink, an invisible design covered over in the palimpsest."[6] A metaphor continues to be lacking: and it ought to be that impossible one by which meaning finally reaches its own abode and becomes "proper meaning," truth, presence of itself to itself, identity. The turning of metaphors, their circular specularity, is this tropism toward the proper, which if it had a result would confirm the very end of the metaphoricity.

What, then, are the histories of philosophy? An obligatory game of infinite and apparently free deferrals: the always renewed promise that coincidence with the proper is only temporarily abrogated. That one day—not yet today—the stone will appear for what it is, a pearl of light. This history is a proceeding according to a false movement of progress: toward reappropriation, presence, self-presence, identity. The Cartesian gesture is the culmination and the modern beginning of this journey: the traveler—that is, ourselves—will believe that he is choosing the route and will not realize that he himself is the effect of the journey. The black light is effaced in the pallid writing with which every metaphor seems destined to write itself, in a circularity without end, without difference, if not that small and illusory deviation with which it slips from figure to figure, never effacing the fabulous scene in the same way and once and for all. But Derrida does not linger, not here at least: the epitaph for metaphor that he composes is so grief-stricken that it urges him on to an almost detached observation. Only the enigmatic "finale" of "White Mythology" introduces a spacing into a discourse that nevertheless takes on the tone (not apocalyptic)[7] of a demonstration. The heliotrope, the sunflower, is an unrealizable homonym: whether one deals with Plato, Hegel, Nietzsche, or Bataille, this flower always brings with it both its double and its end, a dried flower forgotten in a book. No language can escape this repetition or eliminate from itself "the structure of an anthology. . . . Unless the anthology is also a lithograph. Heliotrope also names a stone: a precious stone, greenish and streaked with red veins, a kind of oriental jasper."[8]

How can the doubt that metaphor is also a precious stone insinuate itself within a register of exhaustion in which "meaning" returns every time to belabor the same point, like a death knell? What does "precious" mean if this stone is the weight from which language cannot free itself? How can a "meaning" be "precious" if, according to Derrida, we have in every way to try to get around it, keeping damage to the minimum?

Intonations

Faced with the recognition that metaphor is evasive we can position ourselves in two different ways, we can adopt two tonalities of discussion. We can think that this "escape" is a circle whose limits are already marked out and then show the circle as a cage from which philosophy would continually like to free itself without realizing that this continually reproduces and legitimizes it. Or we can think that in this deferral something escapes us and hence position ourselves in order to try to catch the strange light, the effect as of a precious stone, which surrounds the gesture of fading and flight from oneself.

In the first case the tone would initially be one of observation: a hypothesis must be unfolded. We would study philosophers—Descartes or Husserl, for example—with the discreet distance of the analyst who scrutinizes them while they are tackling with metaphor and would like to free themselves of it. Derrida's reference to a passage of the *Entretien avec Burman* is significant. Now the Cartesian gesture—about which "White Mythology" opens a parenthesis—seems entirely inscribed within rays of the natural lumen, "aether of thought"; at the end of the *Third Meditation* the scene is lit by the "immense light" of God, which Derrida seems to understand (very differently from Lévinas) as an amplification of natural light.[9] Derrida emphasizes that Descartes carries out the discourse of natural philosophy, certainly not that of theology, and in the conversation with Burman we read the following: "The narrative of creation [in Genesis] is perhaps metaphorical; thus, it must be left to the theologians. . . . Why is it said, in effect, that darkness preceded light? . . . And as for the cataracts of the abyss, this is a metaphor, but this metaphor escapes us."[10] Derrida catches Descartes in the double movement typical of metaphysics: the referral to metaphor (which is the very figure of referral) as the hidden source of light, of meaning and truth, coincident with the need to escape from metaphor, filling the gap that the theologian leaves open in order to grasp what escapes us. A descending movement (a return) inscribed within an ascending movement (the process-progress of natural reason). Descartes encounters metaphor as a disturbance of natural light. And Derrida is eager to point out both that metaphor is not a disturbance but the other side of metaphysics and that Descartes, like every philosopher of reason, considers the metaphoric as a weak and confused moment, something that is not yet a knowledge. The result is that metaphor is distanced as a movement internal to metaphysics and that the black heart of the light—here its hidden source—proves to be completely reabsorbed in the play between presence and absence—that is, in the very circle of metaphysics.

In paragraph 36 of his *Lessons in Time*, Husserl confronts the problem of subjectivity as temporal flux: but in the very moment in which we say the word flux, he observes, we are already in the presence of something that is temporally objective. Language has already made us lean toward a constituted fact whereas we wanted to grasp a constituent phenomenon. The "absolute subjectivity" only lets itself be indicated "with an image," with the metaphoric image of flux: "In the lived experience of the present we have the point, original source, and a continuity of moments of resonance. We lack names for all of this." In *Speech and Phenomena*, Derrida does not let Husserl's conclusion escape him: "Names fail us."[11] It is an admission that would reveal how phenomenology can only remain caught within the "metaphysics of presence," notwithstanding that lived experience encounters alterity within itself and that filling identity with meaning is deferred in a process that is infinite. Instead, the historical passage from Platonic *eidos* to Neo-Kantian and phenomenological *eidos* only contributes to showing the obligatory play between light and shadow, between metaphor and metaphysics. It is the difference which produces the subject, Derrida affirms in a Heideggerian tone: difference may be discovered within the horizon of subjectivity, but if we remain within this horizon difference can present itself only as lack or absence, correlative to presence; in this way presence (self-affection, lived experience, the acting identity) will reveal itself in all its metaphysical movement. "Husserl continually warns us against these metaphors."[12] And he cannot do otherwise because "names fail us": he has to admit that the essential (the proper) of subjectivity is unnameable, only substitutable with images (figures) that do not fill but instead underline the lack.

Derrida considers Husserl the philosopher who strenuously looks for "purity," who continuously warns us against metaphorical impurities and, nevertheless, does not find names and has to content himself—precisely in the most decisive areas of his thought—with imprecise images. Here as well we perceive a distance between the observer (Derrida) and the object of observation (the metaphor): the tone of the demonstration closes the task, marking the outlines of an already evaluated repetition. But then, what did Derrida mean when, with regard to the *cogito*, he compared the Husserlian gesture to the Cartesian one, speaking of the impetus and madness of reduction?

Within the same interpretive horizon ("the Husserlian theme of the living present is the profound *guarantee* of *meaning* in its certainty"), another tonality rang out—the same one that echoes in the metaphor of the black light. Just as Derrida was able to show the mad audacity of the Cartesian gesture which drives it on into a territory "so little natural," so one could show the metaphoric node to which the Husserlian gesture gives access. Precisely the philosopher who warns against metaphors exposes himself to the checkmate of his project by exploding the living present in an impossibility of naming. But is it really a checkmate? Or is it not instead an opening, a line of passage of a gesture (the "epoché") with which its "impetus" and its "madness" repropose the enigma of meaning which the black light—to which Descartes was able to draw near—entrusts to us as a question?

In his essay on Lévinas Derrida quotes Borges: "Perhaps universal history is but the history of several metaphors." And he comments: "Light is only one example of these 'several' fundamental 'metaphors,' but what an example! Who will ever dominate it, who will ever pronounce its meaning without first being pronounced by it? What language will ever escape it?" Not Lévinas, to whom the polemic is addressed, who speaks of the "face" as apparition of the other and is therefore himself taken by the metaphorics of seeing.[13] Language is metaphorical and the metaphor is always, in a certain way, luminous. But now Derrida's tone is that of proximity: it is not a question of closing the discussion but of opening it, starting out precisely from Borges's sentence. "Light perhaps has no opposite: if it does, it is certainly not night. If all languages combat within it, *modifying only* the same metaphor and choosing the *best* light, Borges, several pages later, is correct again: 'Perhaps universal history is but the history of the diverse *intonations* of several metaphors.' " Light has no opposite: the black light is not the opposite of light (an opposite that belongs to it like absence to presence). It is a question of *modifying* the metaphor, working on its intonations, which mark the history of thought. But what can "best" light mean? Stronger? Even more luminous? No, to be sure: the movement points instead to a contrary direction, one even more complicated to figure.

I have attempted to bring out not the contradictions, or the successive modifications of perspective (the texts referred to in fact belong to a single phase of thought), but the oscillations of tone that I seem to discover in Derrida's writings: they indicate the difficulty of maintaining a position. The difficulty is of like nature to the problem: the revenge of natural light, the "repatriation" of philosophy, no longer precipitous but now more subtle and addressed toward an abode already proportionate to the difficulty yet always necessary. This difficulty plays at Derrida's own expense every time the risk of the exercise rests in the application of the formula: oscillation concerning the value to give to the metaphorical— now a flower destined to wither between the pages of a book, now again the reflection of a precious stone that transforms sight. Perhaps, however, we can attempt to interpret this pendulum: the different positions, although comprehensibly unstable, permit an identification. Passing from one to another the subject is transformed. It is not cancelled but changes attitude. To think that difference produces the subject means taking up a position without thematizing it as one's own and hence not risking the experience—also linguistic—of this attitude. This is one tonality: to identify oneself with the Cartesian gesture to the point of its hyperbole, risking nonmeaning and the impertinent ambiguity of a metaphor that reveals my own impotence to say: this is another tonality of the same theoretical gesture. Between one tone and the other a proximity plays. Is the passage from the most distant to the nearest only a question of degree or gradation of sight, or does a displacement of levels take place in it, a transformation of the gaze? Between the two intonations there flows a transformation of presence to oneself, a movement of identity. It appears like a loss, an eclipse, a fading, an undermining, an erosion. But what are we talking about? While we try—with words that fail us—to decipher an unheard-of place beyond ourselves, we are, in reality, trying to describe a movement that is our very own gesture.

The Black of the Earth?

In the metaphor of the black light, the high and the low, the sky and the earth do not oppose and annul each other. The light, the high, the sky, the "aether of thought" are not overturned into their opposites. Light has no opposite, and it is certainly not shadow: this means that thought cannot dialectically negate itself as thought and that thought of the shadow, of the night, of "madness," is always masked light, an astute reason. Let us recall Blanchot's insistence on the term "neutral": neither the one nor the other, neither light nor shadow.[14] But it is still a beyond. Something else: does the "madness of the day" introduce another scene? One cannot escape the "day," Derrida suggests. In order not to annul themselves the sky and the earth have to fuse: the scene maintains a glow, a paradoxical luminosity that is "nourished" prescisely when it ought to fade away.

To the "black sun" of mysticism Gaston Bachelard,[15] in his reveries of the earth, opposes the *noirceur* of matter which permits whiteness to shine forth: from the *nigrum nigrius nigro* of the alchemists throughout the entire imagery of literature down to D. H. Lawrence, who says of the sun that "it is only its coating of dust which shines. . . . The sun is dark: its rays are dark."[16] Derrida himself refers to these pages by Bachelard on the imagination of a black, material heart of light (the black interior of the swan, and Cocteau who writes, "The ink that I use is the blue blood of a swan"; or the black milk of the nocturnal goat about which Rilke speaks).[17] The black sun of the mystics (for example, in the "visions" of Saint Theresa of Avila) was a blinding due to too much light: the eye that received it had to close in order not to be dazzled. An excess of light corresponded to the black spot of vision.[18] Bachelard moves us, instead, into the depths of the earth, within our own material covering, which we imagine as being dark and mysterious and which, nevertheless, is the heart, the abode of things and therefore of ourselves, the place of "rest." From the sky to the earth: an imagination more perspicacious than thought drives us down, in our place, where images take on the power of reversibility: an oneiric world of matter whose qualities are hidden.

Derrida, however, follows Bachelard's suggestion only in part. He refuses the game of inversion, which still appears sustained by a dialectical movement: shadow as the inverse of light. Instead, the two experiences present themselves together: it is a question of limit. The problem is to identify the maximum limit to which the absence of light can be forced. The madness of the day is not the epiphany of a dazzling moment but remains a blinding. Earthly experience—yet no cavern of the unconscious offers itself for exploration, no inversion in the depths occurs.

This blinding is not the world of an outside or an other. The Cartesian gesture is not arrested in the face of something that is greater, that cannot be controlled. In fact, it is not a gesture of conquest but of erosion: it is an attempt to undermine oneself by weakening natural light. The blinding is the ultimate experience of an exercise in which the subject proceeds against the grain. We might say that the subject negates itself as self. That is, it negates itself as ability to see itself fully and, therefore, to control itself. But at the same time and following this same path, it searches for itself. The black light is a metaphor of the subject: it suggests its movement of erosion which, in order really to be such, cannot but tend toward a hyperbolic and blinding limit.

Phenomenology and Metaphor

Within this perspective it is interesting to return to Derrida's critique of the phenomenology of Husserl and Lévinas. The metaphor of the black light can be interpreted phenomenologically as a metaphor of the subject. Derrida's oscillations take us back to the "madness" of the *epoché*, whose tendency toward hyperbole—to blinding—cannot be blocked because from the very beginning it reveals itself as an essential characteristic of this movement. We have ended up in the phenomenon of subjectivity: to say that difference produces the subject is a way of translating conceptually—with detachment, in other words—the risk of a gesture that does not aim at an "object" but that interrogates itself as the impossibility of objectifying itself. The discovery of absence in presence, of the other in the same, and of difference in identity has two moments: in the first the play of reference is maintained within the illusory state of an exchange between light and shadow, governed by the light of presence; and in the second it is the very light of presence that is undermined.

Stopping at the first moment, phenomenology would be a modern form of metaphysical closure. But can the *epoché* stop? The phenomenological tone of the two moments consists in the movement, in the gesture itself, in the vicissitude of the *cogito*. There is no other scene: when we evoke it we make use of a disguise, and usually we effect a step backward. This is why Derrida—and not paradoxically—defends Husserl and criticizes Lévinas's "metaphysics of alterity." Intonations of the same metaphor: declension of identity. The audacity of the reduction—the Cartesian gesture, which Husserl relaunches—cannot limit itself to the first moment: the rational project of a reappropriation of meaning contrasts with the radical subjective exercise. The phenomenological gesture is opposed to the "enlightened" task of Husserl.

In this way Derrida helps us to see a decisive laceration in Husserl's phenomenology: a dramatic fracture between exercise and project. But it is from here that the most important philosophic effects of this thought are derived: the double and contrasting direction of phenomenology is not an obstacle to be removed but the heart of this philosophic gesture. Husserl's refusal to consider the theoretical importance of metaphorical language—and precisely while the development of phenomenological research is accompanied in his own writings by an intensification of the use of metaphors—reveals his lack of awareness with regard to this necessary and productive laceration of thought; here Derrida's criticism is right on target. The erosion of identity is a metaphoric modulation, a displacement toward a metaphor whose tone is better able to refer—but at this level referral is description, rigor, and clarity belonging to metaphoric capability itself—to the coming and going of subjectivity, its movement of excess and repatriation. The black light points out this excess to us, and at the same time, given that it remains "a" light, also indicates the necessary repatriation. It is necessary in order that the self remain such and not illusorily dissolve itself, in order that there be word and language and, therefore, philosophy. But the movement, like metaphor, is not separable into a before and after: the mobile boundary of excess and repatriation, a single experience, is the secret of the gesture itself.

For Lévinas the question is that of a condition whose metaphorics can instead be borrowed from the relationship between sleeping and waking: he talks of "insomnia" and of "awakening," attributing the same tonality to both figures.[19] Derrida's critique (which does not cancel Lévinas's influence, however, still present in Derrida's most recent writings)[20] is addressed to a thought that would like to free itself from the "light" (the beyond of being, in Lévinas's sense) and in this way commits itself to a boomerang effect, to a withdrawal on this side of the same, modern metaphysical results. To free oneself from the light would mean allowing oneself to be inundated by an external and superior sun, that alterity which Lévinas does not hesitate to call God. It would also mean not comprehending the stakes for which the metaphorical plays, precisely when philosophical language—as Derrida points out[21]—is being transformed, in an important way, into a metaphorical language.

This critique is well-taken and undoubtedly poses a problem to Lévinas's thought. It is, however, also a critique that simplifies the richness of his thought and does not interrogate its philosophically most dense implications. The perspective here suggested should allow us to glimpse this lack. Lévinas's philosophical gesture can, in fact, be comprehensively interpreted as a radicalization of the phenomenological movement in the direction of a key notion—that of passivity. Husserl, too, moved in this direction when he hypothesized a passive hinterland of perceptive experience: especially in the later Husserl, the theme of passive synthesis, already determinant in the *Lessons on Time*, becomes the very place of subjective identity, anterior to all judgments and all categories of the self. So much so that the transcendental and passivity are knotted together, in an apparent contradiction, in the central ideal developed in the *Crisis of European Sciences*—the "Lebenswelt." Lévinas takes up this theme again and explodes all of its philosophical consequences, which Husserl had only cautiously touched upon.[22] First, he excludes the limiting definition of receptivity: passivity is not the contrary of activity—it implicates the entire subjective attitude. But the most significant modification consists in characterizing the movement of erosion and the search for identity by way of passivity. From this point of view Lévinas's thought is, once again, an attempt to describe the Cartesian gesture, letting the accent fall, however, on the most important aspect: the movement itself. The exercise characterizes itself as a step back, a withdrawal. Passivity is the movement of the weakening of identity toward an ethical identity which we certainly cannot consider diminished. The tone of the exercise and the disposition that we manage to maintain toward both things and ourselves are ethical. The truly passive element of passivity, the recognition of the dependence and weakness of the self, is the grafting of a process—which Lévinas calls "retro-descent"—that characterizes the overall tone of subjectivity: thus the ability to maintain oneself *inside* this movement is ethical.

From the very moment in which Lévinas defines the subject as "a passivity more passive than any passivity," the description can only leap from metaphor to metaphor. The term *passivity* already reveals its metaphoric nature. Insomnia and awakening are two of the various figures adopted by Lévinas: they should be taken together because insomnia gives us the idea of an experience of passivity without vigilance, whereas awakening suggests a passage from a nonvigilance to a mode of presence. The recourse to metaphor and to the sliding from one metaphor to another imposes itself upon Lévinas so as to maintain passivity, vigilance, and presence in an experience which none of the three terms, taken alone, could indicate and in which all three, united, lose their proper and usual sense.

If Lévinas opens up the path toward a thematization of the most important aspect, that is, the mode or the tone of the subjective movement of the erosion of identity, and if this precious indication—even without thematizing it—also shows us the unavoidability of the problem of metaphor, then Derrida's metaphor of the black light also maintains a supplement, a further problematicity: the repatriation or the impossibility of getting away from the light. In this particular light, his critique of Lévinas remains valid. The risk of the gesture, in fact, is a double one: one may lose oneself in an illusory "beyond," in a vain madness without day or in an externality that immobilizes us: or one may underestimate the necessity of repatriation, the necessary compromise with natural light. The second and apparently lesser risk puts into question the very vigilance of the philosopher. Not wary enough to be able to maintain himself within the precariousness and sliding of the movement, the philosopher will find that he has constructed "pieces" of theory with the tone of observational detachment and in the light of an identity that is not undermined.

TRANSLATED BY HOWARD RODGER MACLEAN

Even the atopical is admissible if it appears introduced with reason—Aristotle

FRANCO RELLA
The Atopy of the Modern

Atopy

"Being rooted in the absence of place." Only in this way, says Simone Weil, is it possible "to grasp, like all the saints, what is length, breadth, height and depth."[1] The absence of place is therefore what paradoxically allows us to "grasp" space in all of its extensions, to capture its specific "reality." It is necessary, then, to remove from the "place" that which renders it such: what makes it situs and protects us within itself, inside the secure perimeter of its confines.

The "de-situated"—and therefore atopic—space is not boundless, however. It contains the limit in itself which no longer passes to its exterior, like a line of defense, but to its interior. In this sense "atopic" is the truth theorized by Florenskij as the space that comprises everything that can erase it.[2]

Socrates is not like all the other sages, precisely because he is atopos.[3] As L. Robin writes: "In many passages Plato has insisted on what is strange and misleading about the personality of his master: it is his famous atopy (*Symposium* 221d), his character that keeps one from knowing how to *situate* such a being within the human categories of common experience; this is why Alcibiades can only compare him to fabulous beings like the sileni and the satyrs." Or better still, to the statuettes of the sileni: containers in which there is a precious content (215a–b, 216e, 221 and ff.), a *secret*, something extraordinary, "a marvelous thing which imposes itself but which is not explained." This is the Socratic atopy that generates the embarrassment of Alcibiades and "aporia of his spirit." In Socrates—Robin continues—there is "therefore a mystery, as there is a mystery in love. Love, too, is atopos like Socrates because both, in their nature, contain 'a synthesis of opposites.'"

The word "synthesis" used by Robin is improper. In fact, it is not a question of synthesis or of coincidence of opposites but of *complexio oppositorum*. The nature of love is such that its *space* is traversed by a limit that at once unites and divides the lovers. Erotic union is, in fact, as Schlegel was to say, the union of the un-unifiable, the real place of difference, insofar as "it is not hate, as the wise say, but love which divides human creatures and shapes the world."[4]

Love is *complexio*: contact and mixing of the diverse; interweaving and intrigue. Schlegel proposed some terms to express this mixture: Witz, irony, and arabesque.

Arabesque

The modern city has no confines but is traversed internally by a plurality of limits. The modern city is an atopic space which, precisely because of its bewildering character, has always been perceived as a *labyrinthine space*. The Italian poet Leopardi, however, celebrated cities precisely because a thousand limits break up the habitual view, the gaze of reason which orders everything into hierarchies and categories. One is thus forced to proceed beyond these limits with the imagination—*with the noetic force of the image*.[5]

In this sense the labyrinth, no longer an infernal figure, a place of horror to be dominated with the ruse of Ariadne's thread and the violence of Theseus's sword, comes to be a cognitive figure. Benjamin speaks of losing oneself in the big city as "an art yet to be learned." Dürrenmatt, in his exploration of the possibles, proposes an image of the labyrinth as a place of happiness which—with its spirals—protects from the law, from the violent nomos. But in order to grasp the change of the epochal sense of a millennial metaphor, it would perhaps be better to find another name for the labyrinth. Schlegel proposed the term *arabesque*.[6]

Massignon speaks of Islamic architecture, in which decoration does not try "to imitate the creator" in the illusion of stable and eternal forms "but evokes him by way of his absence, in a fragile, incomplete, and precarious guise." Thus are born "the intertwined polygons, variable radius circle arcs, the *arabesque*, which is essentially a kind of indefinite negation of closed geometrical forms." Schlegel had gone even further, hypothesizing that the arabesque was an original form of human fantasy: the manifestation of the chaos from which forms originate in what we can only define as *creatio ex nihilo*. This chaos is arabesque. The highest order in which "the teeming of the ancient gods" is perceivable. Here is the supreme order, supreme beauty, in that chaos "which awaits only the contact of love in order to disclose itself in a harmonic world."[7]

The demonic touch of Eros is decisive even in the figure of the arabesque. The duplicity that unites in a single constellation that which is, that which has been, but also that which was and which will be possible—in short, even the interrupted bifurcations of history, the attempts that have not had a result but that survive as possibilities.

Benjamin spoke of the messianic or prophetic force of the historian, capable of reanimating what had been defeated. But Schlegel, by way of his new concepts, had already thought of the historian as a prophet turned backward.[8] Going beyond the metaphor of the labyrinth, therefore, is also this atopic thought of time and history which impels us toward a profound revision of our mental and cognitive orders. In fact, if we think that the real is only one of an infinite number of possibilities, and therefore "a particular case of the possible," then, as Dürrenmatt says, it "is also conceivable in a different way. It follows that in order for us to be able to penetrate the possible we have to transform the concept of the real"[9]—and this by starting from its elements of greatest resistance, paradoxical at the same time, and precisely from the "thing."

The Thing

"The synthesis—in itself impossible—of absolute opposites within a manifestation is the essential sign of the arabesque."[10] According to the young Schelling, these absolute opposites are the "conditioned" and the "unconditioned" in that the general state of the conditioning of the world renders unsituatable the unconditioned—from which, according to Schelling, philosophy begins. Once the unconditioned has been situated, this problem resolved, "*everything* is resolved."[11]

To condition (*bedingen*) is, for the young Schelling, the operation by which something becomes or is made a thing (*Ding*). "That which has been made thing" is therefore conditioned [*bedingt*]. An unconditioned thing [*ein unbedingtes Ding*] is an unthinkable paradox, that is, a *non-thing*: it is, in Fichtian terms, Ego.[12] Novalis gives a more advanced answer to this problem. The thing is a paradox; in other words, it is a figure in which that dissent brought to light by Schelling remains productive. In fact, "we look for the unconditioned and we always find only things," which have the ability, however, to modify their borders through the demonic touch of Eros, insofar as "every thing worthy of being loved is an object (a thing)—that which is infinitely worthy of love is an infinite thing [*Sache*]."[13] The unconditioned is no longer the absolutely other, external to the thing, and not even, as Schlegel was to propose—a tendentiously infinite combinatory process: it is in the thing itself, understood by Novalis as *Sache* and as *Ding an sich*, as thing in its own right, and as thing in itself.

The thing is therefore the place of a paradoxical synthesis in which both Polemos and Eros act. It is a product but there is "pleasure in producing. Every operation is therefore a polemical operation. Pleasure of synthesis." Love for the thing, precisely because it is a *polemical love*, is, then, "de-construction and new creation of the world."[14]

The mature Schelling was to resolve this antinomy in his *Philosophy of Revelation* by affirming that "things are only particular *possibilities* individuated in the infinite, that is, in universal strength."[15] But the most radical reflection remains that of Novalis. To act upon the thing, Novalis affirms, prophetically announcing van Gogh's painting, means "representing what cannot be represented, seeing the invisible, feeling what cannot be felt."[16]

Van Gogh has a singular relationship to things: he de-objectifies them. By penetrating with his gaze within the limits of the object, he breaks its state of "being conditioned," he changes the status of the object. The conditioned impeded us from seeing the unconditioned in the thing which now, instead, shows itself: the light of the invisible inside the visible. The thing, then, yields to our gaze everything that had been accumulated and secretly guarded within it. The potatoes of *The Potato Eaters* (1885) emanate not only the earthy color of fatigue and hunger but also a light which is much more alive than that dispersed by the blind lamp hanging from the ceiling.

Is Cezanne's relationship to things equally tense and revealing? Think of the persistence with which he painted Mont Sainte-Victoire dozens of times, almost as if the mountain itself might, at a certain point, become light and disappear like a cloud—or of the elbows or the apples which sink down onto the surface of the table in his still lifes, suspended in a possible, infinite slipping only by their extraordinary weight. This is truly a strange world, a world in which an apple can weigh more than a mountain. It is a world in which a revolution in the relationship between subjects and things is about to take place. The change takes two paths. Things are not immobile but are themselves metamorphosis. It may be metamorphosis in putrefaction and hence a dissolution into nothing, as in *Fisch am Strand*, which Oskar Kokoschka painted in 1930; or it may be *Forms in Combat*, painted by Franz Marc in 1914 in order to free things to a new life.

In the contemporary scene I think, on the one hand, of the petrification of things that seem to be torn from the museum of the avant-garde in order to be frozen in another museum; and on the other hand, of the changing landscapes, the mobile horizons of certain paintings by young artists:[17] states from which things emerge, that offer us new profiles, a different experience of the world.

Contemporaneity

"May the reign of the poet be the world located in the focal center of his age," Novalis wrote on the threshold of modernity, which, precisely with Schlegel and Novalis, began to present itself as a category of thought and of the spirit.[18]

No one was more faithful than Balzac to this "center," which he called *contemporaneité*. In the energies and forms of his age—ranging from technology to science and fashion—Balzac saw a spiritual density that made that "inflammatory existence" the place of a true *gnostic* experience. The traveler of "contemporaneity" moves with means that "have modified the laws of space and time," in a space that is made up of houses and things but that also presents itself as "an ocean," as "a splendid charge of intelligence." In this world "no harmony of sounds is absent. Here one can hear the hubbub of the world like the poetic peace of solitude." In *Le chef-d'oeuvre inconnu*, *Gambara*, and *Séraphîta*, Balzac theorized that in order to grasp the *harmonies* in this apparently uniform world of sounds it is necessary to develop a thought and a poetics of *dissonance*. This world, in fact, has to be *atopically de-situated* with respect to the laws of habitual representation in order to see real harmonies in it, the correspondence between things that always speaks the language of difference and dissimilarity.

The texture of correspondences is in reality foreign and familiar, atopical and bewildering, says Baudelaire, who inherited the theme of "contemporaneity" from Balzac and developed it into a true *theory of modernity*.[19] Surprise, risk, adventure, and peripeteia "vaporize" the habitual Ego and lead the subject before the great paradox of an identity that is always constituted by the other, in a relationship in which the greatest proximity is also the discovery of the greatest otherness. And if Solov'ëv and Lévinas were to acknowledge this paradox of proximity-otherness in love,[20] Baudelaire had already given it a dramatic and "gay" representation in the poem entitled "A celle qui est trop gaie." The poet draws close to the lover-sister. But the habitual relationship of love is not sufficient. In the "flanc étonné" he wants to open a "blessure large et creuse" in order to infuse "à travers ces lèvres nouvelles, plus éclatantes et plus belles" the venom that makes them more than lovers.

The Ego recognizes the extreme foreignness even of what is loved. One can only have the extreme experience of the continual mutation of the other and of things while the Ego also mutates with the other and recognizes itself only in the bewilderment of the transformation. This is perhaps the fundamental experience of modernity articulated in Baudelaire's text.

Some years ago Lyotard tried to describe modernity by starting out from a condition that he defined as postmodern.[21] Modernity, according to Lyotard, is characterized by the domination of "narratives" which can be summarized in the great progressive narrative that is identified with the dominant philosophy of history. Without considering that the very notion of narrative is de-situating and atopical with respect to philosophical discursivity, Lyotard ended up proposing as characteristic of modernity all those elements that show "in the modern the unpresentable in the representation itself":[22] in other words, precisely the antithesis of the progressive narrative, which constitutes one of the poles of the *paradox of modernity*.

The proposal was ambiguous but suggestive. It was received as a liberating voice by theoreticians of the pictorial and architectural avant-garde, understood as a pure and simple territory, historically defined, to be freely crossed. It was also welcomed by the heirs of the philosophy of "decline" and the "end": both prisoners of that historicism which Lyotard had declared typical of the modern—as did Habermas, who claimed the actuality and vitality of the "modern project," identifying it with the unexhausted project of the Enlightenment.

Lyotard has recently corrected his position by declaring that the postmodern "decidedly forms part of modernity" and that "a work cannot become modern if in the first place it is not postmodern." Postmodernism

understood in this way is not, as Gianni Vattimo has said, "modernism at
its end," but modernism in its nascent state—and this state *is constant*.
With a theoretical torsion with respect to his previous position, Lyotard,
then, affirms that "*postmodern* must be understood in accordance with the
paradox of the future (*post*) perfect (*mode*)."[23]

The thesis is once again suggestive. I myself am convinced that there
is a tradition (and therefore a past, an anteriority) of the modern which
has to be (now, in the future) effected. Schlegel and Novalis, for example,
face the same themes as Hegelian philosophy, but they propose a different "visibility" of the world that has not yet been explored. Their thought,
defeated by the Hegelian dialectic, cannot be read within its conceptual
framework, however. It is like an emergence that has moved karstically
along modernity and that today is reproposed in the definition of a different horizon of meaning that involves a complex change of the figures and
concepts to which we have consigned the handling of our cognitive and
operative *situation*.

The bewilderment of the Ego, the metaphor of the labyrinth, the displacement of the light-darkness dialectic in the Kafkian proposal of shadow as place of a different visibility of the world,[24] together with the displacement of the relationship between the real and the possible, are all
certainly included in the modern, but they also—in the moment in which
they become our thought today—work a profound change of epoch.

Today we live in a paradoxical situation in which, together with the
greatest unfolding of cognitive possibilities which man has ever experienced in his history, there is also the possibility—never before so close—
of a total annihilation of humanity and the world. We live in a period in
which what in the past was unthinkable, owing either to distance or to its
dimensions, is today rendered visible on a mass level. After centuries of
interrogating the value of the image in relation to its referent, we are now
confronted by images that have no object-referent whatever. Traversing
the modern really leads us to go beyond its limits, even if these limits are
not external but rather an internal frontier. More than ever, to think the
modern is *to think the limit: it is liminal thought*. From this point of view, Balzac, who sets about investigating the limit between the male and the
female, is prophetic with respect to our present-day thought (he is *after*
our modernity: according to Lyotard, he is totally postmodern). Liminal
thought, precisely because it is poised on the point at which the visible
and invisible touch each other, where place and nonplace are tangents,
is atopic thought. Atopy is perhaps the fundamental word of contemporary modernity.

Secret

As Kierkegaard wrote; "Yes, I think I would even abandon myself to
the devil if he were able to show me every abomination, every sin in its
most horrible guise: this attraction, this taste for the *secret* of sin."[25] The
emphasis here is not on sin, as one might think at first, but literally on

secret, which in Kierkegaard's thought becomes a philosophical and religious category.

The secret is something to discover, but also to safeguard. It produces the pleasure proper to the search, the typical pleasure of the hiding place, as well as proximity to the figure of "horror," amazement, and bewilderment. It is once again Plato who in the *Symposium* proposed this aspect: Socrates hides a precious secret, something extraordinary and marvelous, within his grotesque container. And as Robin writes, it is from this that arises "the inextricable embarrassment, the aporia" of Alcibiades, because in Socrates there is a mystery that is revealed but also hidden in the folds of his figure, as in those of his discourse.

The perception of this "secret" and the impossibility of completely solving it have led some to hypothesize the "existence of unwritten doctrines: the dialogues which our culture has handed down for centuries as a venerable legacy perhaps do not deal with Plato's true concerns, which were instead transmitted by oral tradition."[26] Socrates is consequently a container that contains a stupefying and at the same time disturbing treasure. Perhaps Socrates contains something that does not belong to him. Perhaps Socrates is a character or a pseudonym of Plato, a character who has such a great force as to cancel its author.[27]

The first Platonic dialogues faithfully follow the life and thought of Socrates, but with the narration of his death in the *Phaedo* Plato decisively arrives at the formulation of his own thought, beyond Socrates. And yet, like Conan Doyle with Sherlock Holmes, Plato does not succeed in definitively "losing" his character and speaking directly. He disseminates traces to make us understand his secret, to the point of weakening Socrates' presence in the late dialogues and completely freeing himself from Socrates in his last work, *The Laws*, "making himself the champion of laws that prescribe the death penalty for whoever criticizes or subverts the authority of the country's laws, both as regards the gods and as regards civil order. And so the greatest disciple of Socrates . . . ends up by embracing the point of view of Anytus and Meletus, who had asked that Socrates be condemned to death precisely because of his independent attitude vis-à-vis the then-existing religious and civil order."[28] The secret of Socrates is perhaps the emergence of subjective and *unavowable* instances in Plato, who had nonetheless proposed to "wither" those parts of the soul inclined toward emotion or the expression of the subjective *pathēma*.

Kierkegaard, who meditated at length on the relationship between Plato and Socrates in the *Concept of Irony*,[29] wrote in his *Diaries* regarding the unavowable of subjectivity: "In my papers, after my death, one will not find (and this is the consolation) even one explanation of what has filled my life. In the *recesses* of my soul one will not find that text which explains everything, and which makes events which the world often takes for bagatelles of enormous importance for me and which, even for me, become futile as soon as I remove that *secret note* which is the key."[30]

But the irruption of the secret in the philosophical discourse displaces its rules, *de-situating it* completely. Born from the secret, Kierkegaardian pseudonyms do not limit themselves to safeguarding the secret but give it a narrative development that emerges from the philosophical context in the strict sense: "At the same time as the book develops an idea, its correspondent personality comes to be delineated,"[31] insofar as thought, in having entered the secret dimension of the narrative, can no longer do without the properly subjective dimension. The development of the idea also becomes the construction of a character who lives it and proposes its *experience*.

Kierkegaard, speaking of A *Thousand and One Nights*, reminds us that the narrative is always secretly invested with a vital dimension, "the ingenious intrigue into which the various stories are interwoven between themselves like voluptuously twining plants on the ground, the one with the other. But over the whole there looms the leaden sky of an oppressive anguish: it is Scheherazade who saves her life by telling stories."[32]

Philosophy can no longer pretend that these "names"—the pseudonyms of Kierkegaard, of Nietzsche, and of Kafka—do not exist. And Ricoeur has begun to study the development of the aporias of philosophy within the narrative intrigue, to the point of theorizing a *limit* in which the subject expresses itself *beyond discourse*, perhaps also beyond narrative: in a sort of "meditating lyricism" which opens to a new dimension of thought. In fact, the mystery of time, which pushes one to this limit and is unresolvable philosophically, "is not equivalent to an interdiction which weighs on language; rather, it gives rise to the need to think more and to say differently."[33] It is in this way that discourse, narrative, and tragedy multiply the energies of thought, moving the subject along the path of the unthought. Thus, in a certain sense, the secret is *the soul of the de-situating force of atopy*.

Project

As Schlegel wrote, "The essence of the form of this modern is intrigue."[34] What relationship does intrigue have with the linearity of the project, which is one of the strongest thoughts of the modern?

To project means to construct the place of difference so that what is only possible becomes real. In this sense all of the young Schlegel's speculations concerning the figures of the *combination* of plurality, and especially with *regard to the arabesque*, are the definition of this project and therefore atopical—capable, in other words, of transforming the laws of "only reasoning reason." In fragment 22 of the *Athenäum*, Schlegel confronts directly the problem of the project. "The project is the subjective germ of an object that is becoming," which is characterized, owing to its total subjectivity and its necessary physical and moral objectivity, in relation to time. Projects, in fact, are "fragments of the future." With respect to these fragments "the ability to immediately idealize objects is essential and at the same time *to realize them*, integrate them and partially carry them out in themselves. And, since that which has a relation to the connection or separation of the ideal with the real is transcendental, we could say that the sense of fragments and projects is the transcendental part of the historical spirit." The historical spirit, then, organizes the fragments of the past and of the future as the decisive place of the tension between subject and object in relation to time. It realizes a form which is *a fragmentary totality*, a space of presences and omissions: a place of shadow and mixture.

The criterion of "realizability" advanced by Schlegel tears the project away from the utopian dimension, from pure analysis and pure experimentation, which, as Novalis wrote, "leads into boundless spaces and simply into the infinite," as in "a labyrinth" or in "a delirious *Witz.*" "Realizability" is in relation to what Novalis defined as *finis*—limit and aim—which removes "notorious speculation, discredited and false mysticism," indicating, instead, "the need for limitation—determination, enclosure—[that] refers to a determinate aim and changes speculation into a necessary and properly poetic instrument."[35]

Amazement

The mature Schelling, in his *Philosophy of Revelation*, wrote extraordinary pages concerning the amazement of reason, of stupefied reason, which, faced by reality without limits, without names, without adjectives exists from itself and mutates. Heidegger's pages on boredom in *Wegmarken* and *Grundbegriffe der Metaphysik*, as well as Freud's essay on *das Unheimliche*, are born of this stupefied perception of an absolute transcendence of the real with respect to language which, as Valéry said, suddenly "stops the heart."[36] It is the glance we cast, as Novalis wrote in a fragment entitled "Chance and Necessity," into the "depths of the spirit," when not only reason but also "the power of imagination [lie] *exhausted and immobile.*"[37]

And yet, as Plato and Aristotle said, philosophy originates in amazement. Like the secret, amazement sweeps away "those phantasms that in the sanctuary have" habitually "occupied the place where the statues of the gods ought to be set," as Novalis wrote.

Once again we must return to the de-situating force of atopy. Novalis's move is to use physics for the internal world and the soul for the external world. Normally, he says, "we have stopped at the intellect, the imagination and reason which are the miserable compartments of the universe in us," without a word "regarding their wonderful connections, passages, and representations. No one has thought of going to look for other energies that are new and without *name*. . . . Who knows what wonderful unions, what miraculous generations still await us within the internal."[38]

A wonderful landscape, disquieting and familiar, like the one that Baudelaire hypothesized in the *correspondence* of things. This landscape was explored by Freud, who, with the concept of *Unheimlichkeit*, reintroduced the notion of atopy upon the stage of twentieth-century thought. It was explored by Jung, who saw the demands of *animus* and *soul* move themselves in the subject, and behind them the *Shadow*, the magnificent chaos that awaits the touch of love in order to unfold itself in a world of symbols. It was explored by the great modern writers, from Balzac to Kafka, and closer to our own days by Peter Handke, who, in the *Die Lehre der Sainte-Victoire*,[39] speaks of the *invention of the landscape*, when a flight of a bird or the frond of a branch transforms the dreadful anxiety of estrangement in another place, in which we can once again find the very sense of our journey: *nostos*, return home. In fact, as Novalis wrote, "philosophy, strictly speaking, is nostalgia, the desire to *find oneself everywhere at home*." But if our home is everywhere this means that we are rooted in the absence of place—in atopy, in other words. We have returned to the principal point of the experience of the modern: the journey that transforms the "everywhere of the labyrinth into *one's own* home."[40] It is the metropolitan horizon that takes us through the horror of its *intérieurs* and its dispersion to the place of an unknown beauty that surpasses every canon. Its fullness, as Schlegel said, moves from the celestial dimension to the depths of the infernal abyss.[41] It was described with fear and a shudder of horror by Zola in his great novel *L'oeuvre*, pursued with desperate fury by all the avant-gardes, and today resurfaces in the new landscapes of *our* modernity, of our own contemporaneity.

The subject looks at these landscapes and is in these same landscapes. In fact, one is never only the witness of a thing that happens as we watch it happen.

TRANSLATED BY HOWARD RODGER MACLEAN

FRANCO RELLA
Fabula

1.

    As Valéry said, a great act of strength brought an end to the controversy between Aristotelians and Neoplatonists at the beginning of the seventeenth century and marked a turning point in the history of thought. This was Descartes's gesture, his affirmation of a pure object of thought which brings exclusive cognitive responsibility upon itself. The Aristotelian cognitive paradigm was crumbled, the Neoplatonic world was deprived of its soul: the great body of the *Timaeus* became a surface crisscrossed by measurements, defined in quantities and sizes. But this same process of de-animation (or, to put it better, of "de-animalization") must also be directed within the subject in order to free it from the illness of the body, which, with its illusory force, convinces that being is what appears, how and as it appears: the hardness, softness, its coloredness, etc. Error is always born of the body and extends into intellectual procedures, weakening them, because of the persistent memory of the sensations it gave us during our infancy. To arrive at the ego of knowledge, the ego of the *cogito* which is decided by a pure intellective act, it is necessary to drain "the far too yielding brain of infants of the frenetic and the lethargic." In short, it is necessary "to detach the mind from the senses."

    But *this is precisely Plato's objective in the* Phaedo *and the* Republic: against the "enchantments," the "disorientation," and the "deceptions" produced by the senses, as Plato wrote in the *Republic*, "measurement, numbering, and weighing have shown themselves as being very ingenious aids," operations that "are the concern of the rational element of the soul," which is undoubtedly "the best of the soul," whereas the other, that which trusts to sensation, "will be of less worth." It is "ill" and must be "healed" (602d).

    Descartes's certainty is Plato's *mathēma*: victory over the vertiginous impermanence of the real which ends up by being translated into a theology.

    The Scientific Revolution and mechanism paradoxically annihilate Renaissance Platonism by repeating the gesture of Plato. Giordano Bruno's universe "without a shirt" (as Kepler wrote), the infinite plurality of worlds and astral influences, and the innumerable proliferations of forces in the worldly body, are once again known in *a* form by a subject which is—yet again— "without organs," insofar as "true sight" is not that of the body but that of reason.

Descartes's anti-Platonic gesture therefore once again realizes Plato's cognitive strategy. The anti-Cartesian polemic of a Platonist like Vico opens up breaches within this same strategy. Vico, in contesting Descartes, reopens the ancient controversy regarding knowledge. The notion of a *vera narratio* makes progress by way of these openings: poetic fiction as access to truth.

2.

Vico rallies common sense and the verisimilar against the *scire per causas* of Descartes. But the breaking away that he achieves with respect to previous philosophy is opened up precisely in the consideration of poetry, of the narrative *fabula*, in relationship to truth.

As K. O. Apel wrote, it was a humanistic topos to attribute poets with a preferential path with regard to truth. Even the nonhumanistic German tradition—from Meister Eckhart, by way of Jacob Boehme, to Benedict Franz Xavier von Baader and John Georg Hamann—"drew from the Bible, that is, from the word of God, the vocation of a creative poetic language." But Vico broke away from both these traditions, beginning with the *Orazioni* and the *Scienza nuova prima*: "Falsum poeticum esse quoddam verum metaphysicum, seu, ut loquuntur, cum quo vera physica comparata, falsa esse videantur." Or, as he affirms in the *Scienza nuova prima*: "So distant is it from the real that poetry comes from innermost wisdom! That the false poetics are the same as those which in general are the real ones of philosophers, with the difference that the former are abstract while the latter are dressed in images: because one senses the extent to which he is cunning, he intends it, or to the extent that he is unknowledgeable does not, whosoever writes that the lesson of poets is unbecoming to philosophers; when the truth of poets is a truth in its best idea and the truth of historians is often due to caprice, due to necessity, by mere fortune."

The truth of poets, then, is the truth of philosophers with respect to which it not only is not derived (from the knowledge retained by philosophers) but also enjoys a priority of a historic order (the lesson of poets is not unbecoming to philosophers) and also of a metaphysical order, as it were.

With these opening remarks Vico had already marked out a clear-cut caesura with regard to the humanistic tradition (cf. above, 2.13) insofar as it is the *falsehood* of poets, namely, the "fictional" structure of poetic narration, which is the truth of philosophers. As Vico affirmed in the *De studiorum ratione*, the poet achieves truth *per mendacia*, by way of a fiction. The difference between poetry and philosophy, on this level of reflection by Vico, is posed in that the first "is dressed in images" and the second is abstract. Consequently, "the studies of metaphysics and of poetry are *naturally opposed, the one to the other,* . . . the thoughts of the former are all abstract, the concepts of the latter then are *more beautiful when they are made more bodily:* in other words, the former studies that the learned know the truth being free from every passion . . . whereas the latter attempts to induce vulgar men to strive toward truth with machinery of greatly disturbed effects" (*Scienza nuova prima*).

Behind these affirmations by Vico we can glimpse—as he explicitly recalls—the reemergence of Aristotle and implicitly also of Gorgias. Vico, however, moves beyond both of them. With the *Scienza nuova* of 1744 he arrives at a definition of a true "logic of poetry": the modalities by which poetry poses itself as *vera narratio*, achieving real universality by means of the particulars it embraces on its narrative path.

3.

Giambattista Vico's "new science" is a "critical art" which had been lacking in the "search for the real truth about the authors of nations." The first object of this "science" is "philology," which philosophy "had been almost horrified to discuss" because of the "obscurity of causes and the almost *infinite variety of* effects." Vico's study, however, sets out to discover "the design of an ideal eternal history," and in this perspective both poem and myth allow one to understand "fables as true and severe histories of the ways of the ancient peoples of Greece." Narrated in these fables, in fact, is a truth that could not have been told in another language. And this is even more true insofar as we must admit that poetry consists in nothing other than "in giving sense and passion to insensate things."[1]

Poetry is certainly imitation (216), but unlike the mad mirror that Plato hypothesized in the *Republic*, it does not reflect everything in an undifferentiated manner because "poetic characters, which are genera or *fantastic universals*," constitute the *foundation* of a certain knowledge of the apparent in that "all the particular species are reduced to the genera which resembled them" (209).

James Hillman has observed that Vico's "fantastic universals" anticipate Jung's theory of archetypes. But perhaps Vico's intuition goes beyond even Jung's theory in proposing a poetic and symbolic knowledge that exists apart from the dimension of individual experience. The experience of the subject, by way of the "fantastic universals," can be proposed *as a general knowledge of the concrete.*

This notion, in fact, is part of the more general polemic that Vico conducted against Cartesian positions: the principles of science are not to be sought in metaphysical assertions but in the very modifications of the human mind. In fact, "in the night of thick darkness enveloping the earliest antiquity, so remote from ourselves, there shines the eternal and never failing light of a truth beyond all question: that the world of civil society has certainly been *made by men*, and that its *principles are therefore to be found within the modifications of our human mind* . . . which, immersed and buried in the body, *naturally inclines to take notice of bodily things*" (331)

The language that speaks of this link between mind and body, and of this "taking notice of bodily things," is poetic language, not organized in a "rational and abstract metaphysics, . . . but felt and imagined" by men "with robust senses and extremely vigorous imaginations." Thus the corporeal is united to the divine, the visible to the invisible, for "they imagined the causes of the things they felt and wondered at to be gods" (375).

The poets, in knowing, "created" things; whence their "name," poets—"the same sound in Greek as 'creator.'" Hence they took things to their being, they brought things into being "in accordance with a robust bodily imagination and, because *robust and corporeal, they did it with a marvelous sublimity*, so much so that it excessively perturbed even those who created these things by imagining them. It is through imagination, in fact, that the poet makes things be, and this 'creation' must in the same way trouble both listener and creator insofar as knowing is an *experience* which cannot, in this felt and imagined metaphysic" be separated from the body. The agitation that is at the basis of this experience derives from the sublime: from the divine that shines in the corporeal.

Vico, however, does not propose a poetry that would be a sort of "substitutive philosophy," or even, as will occur in the romantic era, a poetry that is the highest degree of knowledge as ideal intuition. Following the path of philosophy, we could certainly define the "impossible credible" of the poetic myth a "ruse," an error. And philosophy can affirm this *from its own point of view*. But it cannot invalidate poetic myth in its peculiarity in that it is a "fantastic speech" which has its own laws and which, as such, is absolutely true. Philosophy, then, will never be able to liquidate poetry as nonknowledge or as minor knowledge that poses itself in the auroral stages of thought when, as Hegel was to say, it has still not yet attained all of its strength. "All that has been so far said here upsets all the theories of the origin of poetry from Plato and Aristotle down to Patrizzi, Scaligeri, and Castelvetro. For it has been shown that it was deficiency of human reasoning power that gave rise to poetry so sublime that the philosophies which came afterward, the arts of poetry and of criticism, have produced none equal or better, and have even prevented its production" (384).

The faculty of reason that produced poetry was so "defective" that the knowledge it gives us has never been achieved by any philosophy, nor by any successive poetry that could make use of the notions derived from philosophy and that could pose, then, as a sort of "truth" beneath "the veil of beautiful verses."

4.

Having postulated a "felt and imagined metaphysic," Vico moves on to define its logic, which he names "poetic logic."

Etymology is a "history of things signified" by names and by words (22) The "poetic logic" section begins, in fact, precisely with one of these stories, a truly fantastic etymology of the sort proposed by Plato in the *Cratylus*, or in our modern age by Heidegger and Lacan. A story, a significant sequence, like an adventurous peripeteia, condensed within a name, but which from this name passes by way of other names, moving in different languages with an unscrupulousness that perhaps we shall find only later in Lacan.

Vico begins by saying that "logic comes from *logos* whose first and proper meaning was *fabula*, fable, carried over into Italian as 'favella,' speech. In Greek the fable was also called *muthos*, myth, whence comes the Latin *mutus*, mute—for speech was born in mute times as mental language . . . whence *logos* means both word and idea." The course is a dizzying one: logos above all means fable, the Greek *muthos*, which has been transformed into Latin *mutus*, and therefore, given this mutism within itself, logos contains the double meaning of "favola" (tale) and "mental idea." Mythos and logos are, in any case, contiguous—a mixture of mental image, concept, and fable—and from this link we can derive the successive significations of "fact," "thing," and finally of "vera narratio," true narration (401).

Mythology is therefore a "speaking in fables," a narrating that is founded, "as has been demonstrated above," on "fantastic genera." It is *diversiloquium* in that the fables "signify the different species or the different individuals comprised under these genera." But precisely for this reason the *diversiloquium*, the speaking differences, becomes a *veriloquium*, the discourse of truth, in that "this fable was defined as *vera narratio*" (403).

Poetry, then, is narration, and narration, with its movement and peripeteia uniting diverse things according to the genera and fantastic universals, founds its cognitive reason within itself. As Vico himself observed, this discourse is a revolution with respect to preceding poetics insofar as it moves through antinomies and mixtures, through a theoretical hybridization which is proposed as a tensional proceeding. The concept itself of *vera narratio* is an example of the tensional status of Vico's thought. Translated into traditional terms, *vera narratio* in fact means "true fiction."

5.

Tropes are corollaries of this "poetic logic." "The most luminous and therefore the most necessary and most frequent and dense is metaphor. It is most praised when it gives sense and passion to insensate things in accordance with the metaphysic discussed above" (404). Once again, metaphor is a synonym of poetry in that it has the same task and the same reason (cf. 186). This identification is possible insofar as metaphor, according to Vico's definition, which is perhaps the "most dense" and most acute that has ever been given of metaphor, is "a small tale" (404). It is therefore a short story, a little myth, a brief narration, synthesis or composition of facts which in itself unites the heterogeneous. In short, it is a fragment of the more general process of the *vera narratio*.

With metaphor, the body enters the heart of every discourse, of every logos, given that "most of the expressions relating to inanimate things are formed by transfers from the human body and its parts and from the human senses and passions" (405).

It is impossible to speak of anything without the body and "human senses" insinuating their disturbing presence into the discursive network. Thus discourse is the image of the universe within which we move; it constructs an uneven topography of lights and shadows, reliefs and lacunae, abstraction and immersion in the body. The supposed absolute purity of scientific metalanguages is merely a white metaphor, an immense figure that once again narrates the conflict of knowledge, the history of the subject and of subjects involved in this conflict.

6.

"This fantastic metaphysics shows that *homo non intelligendo fit omnia.*" And this is how Vico, encompassing both, defines them. "When man understands he extends his mind and takes in the things, but when he does not understand he makes the things out of himself and becomes them by transforming himself into them" (405). This is not Cusanus's doctrine of the learned ignorance, which also might have influenced Vico. *Intelligere* and *non intelligere* coexist in two different strategies of approach to the world. But precisely this coexistence brings with it a new term. By "not understanding," man is transformed into the thing: a continuous exchange therefore comes about, a continual metamorphosis between subject and things, between the Ego and the world. Mutation thus becomes one of the capital terms of the new fantastic metaphysic. The object of the knowing of narration.

7.

Poetic "falsehood" is truth in narration. Lie, on the other hand, is proper to abstract reflection that "takes on the mask of truth" in some of its linguistic-argumentative procedures. "The first fables," relating the events of the world as they appeared to men, "could not feign anything false, they must therefore have been, as they have been defined above, true narration" (408).

8.

"Monsters and poetic transformations" derive from the cognitive activity that composes "the subjects" in order to make up their complex forms, or that destroys "a subject in order to separate its primary form from the contrary form which has been imposed upon it" (410). Thus "the distinguishing of ideas produced metamorphoses" (411).

What for Plato was evil itself—the "metamorphosis of the immutable" of which poets stubbornly insisted upon speaking—becomes essential in Vico's cognitive strategy. But what appears as a process of metamorphosis is the composition and decomposition of the subject according to its "nervous system," a process Plato describes in his *Phaedrus* when he defines dialectics. When this Platonic metaphor is read as a "small fable," the process may be seen as one of real movement, of metamorphosis. Plato, too, describes a movement in which one image becomes another: a process of transformation and transfiguration.

9.

The conceit of nations and scholars has negated poetic wisdom. Vico polemicizes against the recourse to transrational philosophies, such as those of occultism and hermeticism, because with this "irksome philosophic wisdom" philosophers, pretending to wish to attribute great value to poetry, in reality negate it in its cognitive specificity insofar as they refuse to acknowledge its peculiar logic. And this logic may be understood only within the double movement of understanding and not understanding, within a complex framework in which sense and intellect—by different paths—find a language capable of giving form to the world: in order to bring to their Being those things that otherwise would lie in a confused and lifeless heap (779, 125-28).

10.

The "real Homer" must be reevaluated within this framework and against the Platonic condemnation of the *Republic*. His "inaccessible lies" and his extraordinary cognitive fictions (838) have proposed themselves as a model of wisdom which is still unsurpassed.

'Omēros—Vico tells us with yet another of his striking etymologies, true "narrative fictions" that try to deliver a truth otherwise unattainable by way of "pure" thought—derives from ὅmou, "together," and from *eirein*, "to connect." Our Homer was therefore "binder or compiler of fables" (852).

11.

Vico's strategic moves might have but did not, in fact, resolve the "conflict of knowledge." The battle was to be resumed in the "modern" period and with uncertain results.

TRANSLATED BY HOWARD RODGER MACLEAN

MASSIMO CACCIARI
The Problem of Representation

Only the Angel, free from demonic destiny, poses the problem of representation. Demonic destiny "ist der Schuldzusammenhang des Lebendigen" (Benjamin, "Schicksal und Charakter"),[1] the guilty context of everything that lives. Guilt, for the demon, is the constitution of the living: it refers back to an original guilt that condemned at the first incarnation and set into motion the *rota generationis*. Life itself is condemnation. The *daimon* nails it to the laws of destiny, which have nothing to do with those of justice.[2] The dimension of the *daimon* therefore absorbs that of character: character "abdicates itself in favor of guilty life" (ibid.). As it does in Plato as well: having chosen one's own life the soul is constrained by the daimon. Ethos becomes a demon for man, which he can escape only by way of a new death and a new birth.

Benjamin wants to break the chain that connects the concept of character to that of destiny. Knowledge has always worked to weave an extremely fine web in which the two concepts become indistinguishable wefts—and hence every character can be judged by it in the light of an unchanging Law, according to Jurisprudential Right. Character, in this way, makes itself *daimon* subject to destiny. Within this context a *problem* of representation does not seem to exist, given that character *is* the *daimon*, and the *daimon is* the power of that destiny. In the fabric that links these terms neither questions nor voids appear. But for Benjamin, tragedy has already lacerated it. In tragedy, pagan man tries to summon together his strength "heimlich," secretly; his victory against the *daimons* begins. "Greek tragedy honored human freedom because it made the hero *fight* against the superior force of destiny," as Schelling had already written in the last letter of *Philosophische* Briefe. Blame and punishment are jostled together and confused by the tragic poet: to every Nomos another is opposed; to every Logos there is a counter-melody. The discourses of tragedy appear to be *dissoi*—doubles—and to the greatest extent *dissoi* are those of god, servant-guardian of Ananke. However much this knowledge takes away the word from man, reduces him to silence, and, apparently, annihilates him, it is here that "the head of genius was for the first time raised from the fog of guilt," and he felt himself essentially free from the demonic destiny. Certainly, the hero has necessarily to succumb, and yet, given that he succumbs not without struggling, he shows in his defeat, by way of that very loss of his freedom, that in the defeat there occurs "precisely this freedom": he succumbs "with an open affirmation [Erklärung] of free will" (Schelling, Philosophie der Kunst).[3] The adventure of his autonomous logos also begins here. In comedy, character "luminously unfolds" without allowing anything to exist next to it; comedy affirms the freedom and autonomy of character from the daimon. The comic persona, far from being "the puppet of the determinists," opposes "the splendor of his unique feature" to the finely woven fabric of Destiny, to the primacy of guilt it opposes "the vision of the natural innocence of man." The comic persona *plays* its own demonic character by defining its own irreducible individuality with lucid consequentiality.

But such autonomy presupposes that of logos, of the form of expression proper to the comic persona. Its freedom, that is, as Benjamin notes, without however developing the problem, is affirmed by way of its affinity with logic ("auf dem Wege ihrer Affinität mit der Logik"). The tragic word knows nothing of this affinity; precisely in its agitating together and confusing guilt and punishment, it could not follow the criteria of logic. It does not represent but re-strikes a sound whose origin is hidden, an *adēlon*, an inscrutable.[4] With respect to this *adēlon* the tragic word cannot proclaim any form of autonomy. Instead, autonomy is the logic of comic character, no longer a simple "knot in the net" (Benjamin) but an *individuum* that unfurls itself on the basis of the uniqueness of its fundamental trait or *temperament*. The discourse of this individual becomes substance unto itself, complementary to nothing, corresponding only to its internal order; the "comic" aims at the annihilation of every reality that lies outside the unfolding of character: from the point of view of the concept of destiny and of *daimon*, a vertex of hybris, of arrogance.

Hence the problem of representation explodes, already originating with the tragic form. What relationship does an autonomous logic have (no longer repercussion, resonance, a link) to the thing? In what way can a logos that is metaphysically detached from every presupposition represent a thing that is different and heterogeneous from itself? How can the logoi stand for the thing itself if no common origin is expressible, or rather, if the logos, in breaking the net of demonic destiny, has absolutely and autonomously defined itself? In tragedy, a logos that already radically doubts every stable, transcendent foundation still torments itself in order to arrive at the idea of it and *sinks*, it goes right to the bottom, in this desperate search. Comedy criticizes the judiciousness of this search: character is not a demon, but every demonic dimension is here totally subsumed within the individuality of character. The appearance, the manifestation and expression of character counts for everything. But this expression cannot be other than *onomazein*, the *onoma*, of pure names "which mortals laid down believing them to be true" (Parmenides, frag. 8.39 Diels-Kranz). The name reflects the *doxai*, the opinions of mortals, the deceiving order, the apparent disposition of the world. The name is the arbitrary instrument of men "eidotes ouden," who know nothing, "dik-rānoi," two-headed, for helplessness guides the wandering thought in their breasts (Parmenides, frag. 6 D.-K.). If the logos beats the path for which "to be and to be-not are the same, yet not the same," if it renounces the absoluteness of the *alētheia*, if it erects itself in auto-nomy with respect to the unbeatable force of Necessity, then it is necessarily resolved in the *onomazein*. But it is impossible to make any certain and stable Being correspond to the *onoma*; it is nothing other than the representation of the oscillating of the entity in the *doxa* of mortals. Properly speaking, it represents only the deceiving order of this same *doxa*, in its perennial changing of place and color.

How is the absolutely opposed to the *atremēs* heart of the Aletheia representable? How is it possible to know what never stays? And how, together, is it possible that in the name, arbitrarily posed and perennially changing, one gives *one* representation of anything? The name belongs to the realm of opinion from which, on principle, the truly real thing escapes. Names "are in no case stable. Nothing prevents the things that are now called round from being called straight and the straight round" (Plato, *Seventh Letter* 343a-b). Name, definition, and image form knowledge, but an obscure and instable knowledge of the oscillating entity which has nothing to do with the "fifth entity," with the higher degree of the knowledge of the truly real thing. And if one were to affirm that the *onomazein* may comprise the "fifth entity," certainly he would be afflicted with the madness of mortals, not struck by the *mania* that comes from the gods.

But not only does the abyss of the truly real thing escape the name, not only is nothing known of its *no-thingness* by that knowing which is the ephemeral order of opinion constructed by way of names, definitions, and images—but the very *on* that is sought in the *onoma* (thus Plato explains the term in the *Cratylus*: "on ou masma estin," 421a-b) can never be effectively drawn from the *onoma*. The search that the name carries out is in principle interminable. If, in fact, the image were perfect imitation, if the name stood perfectly for the thing, then there would no longer be either image or name but duplication of the *on* itself. "Do you not perceive that images are very far from having qualities which are the exact counterpart of the realities which they represent? . . . But then how ridiculous would be the effect of names on things, if they were exactly the same with them! For they would be the doubles of them, and no one would be able to determine which were the names and which were the realities" (Plato, *Cratylus* 432d). He who knows the names does not therefore know the things, neither in the sense of knowing the "it self" (439d), the truth of the thing (438e), nor in the sense of a perfect adherence to the appearance of the entity, of a full *touching*, point by point, of the *on*, of a complete re-presenting of it. How can names signify the truth of the thing if they appear intrinsically ambiguous, if they do not agree between themselves, if the legislators who have posed them could most certainly not have been able to avail themselves of other names from which to learn (and so they would not have been able to apprehend the "it self" of the *on* by way of the name)? Is it judicious to "put oneself and the education of one's mind in the power of names" (440c) if the knowledge that comes from them always changes and will therefore always be *non*-knowledge?

The rupture of the demonic character destines this interrogation: *Socratic* comedy. Comedy is the infinite varying of this single question: if the representation by means of the name does not give the thing but the *doxa* around the thing, how will it be possible to arrive at a knowledge of things by way of these same things, at an apprehension of the truth of the thing in itself, of the truth that can always be intuited from a knower who also always is? The freeing of character from the *daimon* coincides with that of the name from the being-signified, that is, from being the imitation-image of the thing it signifies. The name unfolds "in the splendor of its unique trait" (Benjamin), in the solitude of its "temperament." "Free," character roams by means of names and their etymons, combining and varying, a game of intelligence and of invention which challenges the original resonation of the word: does this sense not persuade you, does this explanation seem crude to you?—then see if this other one satisfies you (399e). Can we really *seriously* interrogate ourselves regarding the reason of names? "Ask it of others; there is no objection to your hearing the facetious one, for the gods too love a joke" (406c). A *joke* that Dionysus is the giver of wine, that Aphrodite means born of the foam, Pallas the shaker. The *onomazein* does not "touch" the truth; to learn the thing itself is not possible by way of names. Yet it is that "joke" of the name or of the care for it (which already cast its shadow in tragic drama) that undermines the *daimon* from his dominion over character. And thus the joke of the *Cratylus* appears to be tremendously serious. It is the question that Socrates continuously repeats to himself "as if in a dream": how will it be representable and apprehensible that same "it self" which we consider not when we observe a beautiful face in its passing and never-ever-being, but the beautiful always as it is? How will the idea be representable if, as it turns out, it cannot be apprehended by way of the kaleidoscopic game of *onomazein*?

This problem constitutes that "Angel with the blazing sword of the concept" which Scholem sees rising up at the entrance of "paradise" of the written: the *erkenntniskritische* Vorrede of the *Ursprung des deutschen Trauerspiels*.[5] The angel watches over not only this piece of writing but also Benjamin's entire opus: an opus then, marked by confrontation with the Platonic problem of representation, disturbing the demonic character forced into the chains of Necessity. It is to the "Frage der Darstellung" that, at each turning point in time, at every *krisis*, "mit jeder Wendung," philosophic literature must return. If the reflection of "normal time" forgets this consideration, does not make room for the consideration (*Rücksicht*) of representation, the philosophy of the time of *krisis* or, better,

of time as *krisis*, it takes on as its own essential task, as *its* own, that look-
ing-once-again, the return to meditating the question of representation.
The time of crisis coincides with the time of the return—of the anamnesis—of philosophy as question of representation.

The representation we are dealing with here is the representation of
the idea. Benjamin explicitly affirms this: "If representation wants to affirm itself as an authentic method of the philosophical treatise then it
must be the representation of ideas." But the idea, insofar as it is "property" of consciousness, is not the pro-duct of the intellect's spontaneity;
names are the properties and pro-ducts of consciousness. The idea must
be considered as "ein Vorgegebenes," a something already-given to consciousness. The struggle for representation is renewed at every *krisis*:
when, in other words, either the dimension of the idea as a *Vorgegebenes*
threatens to remain prey of a negative theology (there is no possible representation of the idea), or—but the two aspects can be closely intertwined
—one presumes to be able to represent it by way of names-definitions-
images. When this risk is very great, the struggle for representation as
representation of ideas is renewed, "eternal constellations" not determined by intentions.

The being of the idea is a-intentional, distinct from the connections
of knowing, "exempt from mediation." The idea is not the *eidos* or form of
the observation; it is not form of vision. Benjamin explicitly polemicizes
against the neo-Kantian interpretation of Plato. The idea is not the *eidos*
that in-forms the representation. The problem consists in the representative giving of itself of the idea, not of the forms in which a "civilization of
vision" represents it. "Truth is the death of intention"; any theory that
wishes to reduce it to the ambit of the intentional relation is destined
not to get to the bottom "of the peculiar giving of itself of truth, a giving
of itself which evades any kind of intention." The forms of the analytical-
conceptual connections consider themselves representations of the truth,
in the same way that names claim to possess the very thing itself. In reality, this conception only shows ignorance of the problem of representation. If representation here equals an image that is identical to the thing,
we return to the aporia of the "double" of the *Cratylus*. We would then say
that the forms of the connection are truth itself. But the forms of the connection are made up of names and verbs, they form the world of the experience of the entity in its oscillation; thus they are to be rigorously distinguished from the forms of the giving of itself of truth. If it were we who
pro-duced them, the truth would be pro-duct of our intention; but our
intention can only develop itself—once again—by way of names and
verbs. Thus the problem of representation must only be understood as
the problem of the self-presentation of truth itself.

A philosophy that corresponds to this problem does not find itself
in the condition of mere "research," to which a "current conception"
makes it equal or even subordinates it.[6] The researcher moves within the
ambit of simple "extinction of empirics" or the "negative polemic." His art

of confuting only shows the imperfect nature of the "four entities," the unattainability of the "fifth" by way of knowing that originates from the name. He is an ironist. His art "extinguishes" empirics as it shows the constitutive instability of its names, its *never*-being; but in not facing the problem of representation, it limits itself to a negative polemic. Instead, the philosopher questions himself regarding the *positive* representation of the idea. Of course, on the one hand philosophy values research because it is also vitally interested in the "extinction of empirics," but it distinguishes itself from research because in the fever of the negative which seizes definitions and images as soon as they are submitted to meticulous skepsis, in the trouble and effort aimed at removing the dimension of opinion and naming, it remains attentive to the splendor, the gleaming of the real in its a-intentional essence. One should note that for Benjamin this means the exact opposite of an ec-static intuition of truth; here it is not a question of "ascent" to a higher vision by way of some initiatory itinerary.[7] The truth simply *gives itself, haplōs*, not mediated by our faculty of representing but suddenly intuited (for Plato intuition is the faculty nearest to the "fifth entity"). This giving of itself is that of the thing—but of the thing itself, of the self-ness of the thing, in principle irreducible to the network of connections represented by naming-defining. The truth immediately gives itself as the thing in itself and cannot be further interrogated regarding its foundation or its reason. The "fifth entity" intuits the truly real, which can only give itself and gives itself precisely by withdrawing from whatever definition. Naming grasps not the truly real, but functions, relationships, and beings which only in their relation prove to be conceivable. Thus each name gathers in itself more or fewer of the other names (it is only understood in the context of other names: *Seventh Letter* 343a); it is intrinsically mediation with the other from itself. The thing itself, instead, passes through the mesh of definition; it shines in every definition as that which always withdraws from it. It gives itself in the definition as its own indefinable. Yet what, in the definition, together withdraws from it is not at all an absolutely transcendent dimension but, rather, the thing itself, *precisely* the thing, the this-here *individuum* of the thing.

The *thing* must be said to the Angel. Precisely the thing *itself*, "invisible" to functional definition, must be brought to the invisible Angel, to the Invisible that is the non-place of the Angel. But how can it be *said*? Is there a form of saying that is extricable from the *onomazein*? Is there a word by which that truth which has the pure and simple consistency *haplōs* of the thing can be given? The *Vorrede* of Benjamin turns upon this question. The name that for Benjamin "determines the giving of themselves of ideas" cannot be understood as a form of the *onomazein*; it must appear, as it were, *after* the Platonic critique. The name, for Benjamin, excludes any "explicit profane meaning"; it determines the giving of themselves of ideas only insofar as a *symbolic* dimension belongs to it—that is, only insofar as the idea arrives at *self-transparency* in the name. The thing itself and the name thus form a symbol. One must listen to the name resonate in its own right, as "it self," *at one* with the thing; not the name insofar as it *serves* the definition of the thing within the network of its relations but the name as sound of the thing itself, one with the giving of itself of the thing. Just as naming is functional with regard to the representation of the relationships between beings in their flowing, so also the name may appear as the symbol of the thing itself, a-intentional self-transparency of the thing itself.

It is the name-symbol that speaks to the Angel, through us, the most ephemeral ones. The name, in which the thing "saves" itself within itself, communicates to the Angel, to the dimension unreachable by the *onomazein*. One communicates with the Angel by way of the *intransitiveness* of the name;[8] if, that is, an intransitive dimension of the name gives itself whereby the name resounds as the thing itself, without reason or aim, then the idea is representable. For Benjamin, philosophy is essentially the uninterrupted struggle for the restoration of the "primacy" of the symbolic in order to make room for the name as symbol, for only in the name-symbol does the idea give itself. The *Frage der Darstellung* becomes the problem of the name that is abstractable from the *onomazein*, of the in-transitiveness of the name, *hence* symbol of the thing itself, *thus* determination of the giving itself of the idea. The task of the philosopher is carried out under the sign of Eros: Eros for the very representation of the name-symbol, Eros in order to say to the Angel the thing itself by way of the "intact nobility" of the intransitive word. *To say the thing* is (would be) to already tell the Angel, since to say the thing is (would be) to say that *realissimum* which cannot be said by way of the *onomazein* (in which subject and object remain separated), but only by means of a name that is the symbol of the thing—self-transparency of the thing itself as such— "saved," that is, as idea. The name-symbol of the thing that is "safe, at the end," of the thing "as a thing that is" (Rilke, *Seventh Elegy*, 69-71), this *idea* of the thing must be found again, "erected," in the gaze of the Angel. There one must listen to the sound of the still-living word: House, Bridge, Fountain, Door, Window, Tree, Tower, Column. One must say them *in this way*, as if none of "these" things, captured in the net of discourse, has ever been intimately understood to be—say them each as an individual idea. Better: leave them to become transparent in the symbolic dimension of the word which, even though in the "streets of the city of torment |Leid-Stadt|" (Rilke, *Tenth Elegy*, 16), still resists intact with itself.

The name-symbol does not possess the thing but represents its giving of itself. Those words, pronounced by Rilke, have the *consistency* of things; they are not the thing but are like the thing itself. The Angel orients man not to the conquest of that which cannot be detected but to the recognition of his becoming self-transparent in Eros of which this *like* is only the manifestation. But neither is the Angel, as we know, the hermeneut of the highest Point but rather the patient exegete of its infinite names. The pathos that moves man toward the name-symbol is therefore shared by the Angel. The figure of the Angel more resembles that of a companion caught in our own event than that of the hermetic-gnostic Psychopomp. The Angel *follows* man; it wants to be named by man's desire for the name. Benjamin's New Angel is the extreme figure of that tradition in which the Angel ends up by being inextricably involved in *all* the dimensions of our saying, of our various and possible saying, and consequently always appears more kindred to their *catastrophes*, addressed, *entwined* to them. Angel, all the same. Even if he always flashes new in an instant, he is the form of the giving of itself, in the time of the *Leid-Stadt*, of the idea in the name. And the representation, given that it means Platonic "salvation" of phenomena, also watches over the eschatological-Messianic motif, precisely the sound of the Angel. Entwined in our allegorical game, in our *Trauerspiel*, to the point of sharing the same fallenness, the Angel, nevertheless, does not lose the thread of the problem of representation. And only then does the allegorical game become authentic Mourning (*Trauer*): when the problem of representation is taken back to its heart, when—in the Angel—it looks for the symbolic dimension of the name—and *here*, in this search, in this trying-to-say, and not to more anchorages, it is shipwrecked.

But of what idea is the Angel the exegete? The idea that is represented in the name of the Angel must express the inherent eschatological value of each representation. He is, in fact, the exegete precisely in this sense: that he causes phenomena *to emerge* from their appearance, out of the slavery of the letter, diverting them from their immediate presence in order to represent them, re-present them according to their truth, thus finally rendering them *justice*. Whereas the hermeneutic exercise renders *right* to propositions, it orders their connections according to right, it *judges* them, the exegesis of the Angel renders justice-*praise* to the symbolic dimension of the name. But this Angel is the New Angel, turned to the ephemeral: how can its figure really manage to make the thing "return," as a point, to the eternal constellation of ideas? Of what idea can the Angel, as the New Angel, be the exegete?

In the so-called esoteric fragment of Ibiza, the *Agesilaus Santander*, his song (which is singing the *Eternal*) represents by *dividing*. His exegesis begins by putting to the test, *accusing*, creating misery, *penia*, for the absence of the loved one. But it is as if this principle of separation were projected onto the background of that song and then, from within itself, matured an incoercible force of expectation, an invincible patience.[9] Accusing and separating he e-ducates the expectation and patience for the name. The wings of the Angel are the wings of this patience, of this *prosokhē* or attention not addressed to an absolutely separate-transcendent dimension or to an unfathomable mythos (better still, to that *adēlon* more originary than any *mythein*), but addressed to the symbolic dimension of the name, which always gives itself, to the symbolic "primacy" of the word, of *this* word: House and Fountain, Door and Window and Bridge. Separation is not given—in the Angel—if not *at one* with this profane attention, at one with the to a-wait that draws us to the future through the same move-ment of the *er-innern* (of the re-mem-bering) the past, in that the Angel has sung (in the instant, and *for this reason* has addressed himself to us, to the ephemeral) the Eternal.

Thus in the ephemeral the thing can be "saved" in the name. This "weak Messianic force" (Benjamin, *Thesis of the History of Philosophy*) con-ceded to us represents itself in the Angel: not pure difference, not differ-ence as separation, but attentive and patient exegesis of difference. The New Angel can e-ducate to this, for he is not simply ephemeral but is the perfect Ephemeral, the Ephemeral that has in the instant known how to sing-praise the Eternal. He protects this weak force for us, that in the ephemeral this may come about. But what form will this fire of exegesis be able to take on (and that no exegesis will ever put out) in our own age, in this time of interminable expectation or *necessity* of interrogating: the meeting with the name, with the word interrogated according to the pri-macy of the symbol?

The Angel is the exegete of a dimension of *time* to which the weak Messianic force still belongs. The name that we share in our symbolic conversation with the Angel is that of this dimension of time. The prob-lem would not exist if between symbol and allegory a simple abyss were to open, or if "progress" were established between the two dimensions (or a "fall"; it is the same thing), irreversibly leading from the former to the latter. The Angel, however, allegorizes the symbolic insofar as symbol-ically representing the allegorical. Neither mediator nor pontiff, he is the name of the original symbolic tension that unites the infinite difference. Angel, but Angel of history—history, but history conceived in the name of the Angel. In this name, history as continuum, calendar of the always-the-same, permanent passage-transition from present to present, succes-sion of *nyn*, ceases to have value. The time of representation of *ideas* frees itself from that of the "once-upon-a-time" of the "brothel of historicism," of the *Universalgeschichte* adding event to event. *Erkennen* to *Erkennen*. Not that the Angel (as has already been seen) ec-statically escapes from the continuum or "transcends" it—but he comprehends those events as "a single catastrophe"; he would like "to awaken the dead and recompose the broken"; he maintains secret understandings with past generations. To break the chain of demonic character is at one with the crisis of time as inexorable succession. And only at this point does the reason for the struggle between the Angel and the demon become clear, between his weak Messianic force and the "principal argument" of unbeatable Neces-sity (that if the possible is, it cannot but be real; that the possible is only strength *of reality*). In the name of the Angel the idea that it is possible "to detonate" this "argument" is rendered self-transparent, to dash away from the homogeneous and empty time of the continuum, create days—*Fest-tage*—capable of arresting the flowing and re-create it at the same time. To entropy, to irreversible consumption, his name opposes the *ek-tropic* instant. An "eternal" image of the past does not appear conceivable in this time, an image of the past as perfectly-been. The past itself is still *in-securus*, it may light up with hope, it can ask for *justice*. Never, in this time, is the past *vanquished*; the present is never a mere field of victors from whom, as Simone Weil said, justice is always forced to flee.

A *new* time is what the Angel incessantly looks for the *just* representation: present-instant, interruption, arrest of the continuum, *Jetzt-zeit*. Every *Jetzt* can represent it. The term must therefore be connected to the passage in *Baroque drama* where it is contrasted to the mystical-symbolic *Nu*. And yet, the *Jetzt-zeit* does not mark the simple "fall" into the allegorical but is, in the allegorical, memory of the symbol. It is on the strength of this memory that the infinite variations of the *Spiel* are transformed into a *Trauer-spiel*, in a game that is illuminated precisely in the mourning for the absence of the symbol. In the *Jetzt* of the *Jetzt-zeit* the time of every "now" is idea of this memory. However weakly, as "hesitant immobility" as "slight flickering, imperceptible," the *Jetzt-zeit* within its name guards the only "model of Messianic time" conceded to the force given to us as dowry. This force is represented exactly in the "saving" the *Jetzt-zeit* from its immediate profane meaning, *enlightening it*, neither transcending nor sublimating it: profane illumination. In the dimension of time as *Jetzt-zeit* which "exceeds" the mere duration, as moment or instant, the only representation of the idea is given to us, of the *eternal* of the idea.

It is that dimension which Franz Rosenzweig comments on with regard to the end of Psalm 115 (113B), the dimension of that "But victorious" which the living ("But we, the living"), from the depths of their fallenness, can raise in the choral praise of the Living, in the instant of the praise, *breaking* the "scene" dominated up until that point by the idolatry of peoples.[10] The greatest idolatry is the cult of the having-been, of the irredeemable that-was. Against idolatry is raised the cry-song of the living to the Living Being. Only at this point—in the moment of this song—can they truly call themselves living; before they were a succession of moments destined to death, born in order to die. The living recognize and affirm themselves living only in the instant, *periculosum par excellence*, in which they praise the God who is *not* (the peoples ask "where is" God, but if God belonged to some place he would be only the genius or the demon), in which they trust to the Invisible a-waiting, e-ducated by the Angel, the names. The time of this *But*, a fragment broken off from the equivalent continuum of the *nyn*, is cut-out, truly chronos from *krinein*, truly *tempus* from *temnein*. Seeming to absolutely contrast this dimension of time held in the cut-out (the instant between stone and current about which Rilke says the purest possible) is that of the *Aion-Aevum-Ewig*, of the "uncut-out-able" *Hodie*, of the perennial light *Dies*. But this contraposition is abstract, it makes the eternal an ab-solutum from time, it loses the concrete and living polarity of the two dimensions (just as the abstract separation between allegorical and symbolic loses the very idea of the *Trauerspiel*). The great scholasticism (Jewish, Arab, and Christian) did not limit itself to contrasting the *Nunc stans* of the divine with the *nunc fluens* of the creature, but added a third: the *tertium datur* is the N*unc instantis*,[11] the dimension of a *sudden meantime*, so sudden, and yet so *actual* as almost not to be felt as moment of time. A dimension of *perfect* fallenness: the most sudden moment (fallen like the New Angel) has for its name the moment that arrests, that cuts-out the continuum. The real name of the most ephemeral, the name of its *idea*, is N*unc instantis*. The "small door" of which Benjamin speaks is the image of this name. Like every door, it has on two sides which unites precisely while separating, H*odie* and *nunc fluens*, indissoluble polarity, inseparable difference.

The Messianic chance, for Benjamin, coincides with the possibility of representing *this* difference. In Christian theology it tends to be resolved in the triumphant *kairos* of the Event, incalculable and unforeseeable, and yet full, definitive, *state*. For Benjamin it is not given "to go back to" the exile and the separation that have broken into the same sphere of the divine (the Angels narrate it). But in the *Jetzt-zeit* that arrests, that cuts-out *is-a-turning-point*—one may recall the splinters, sparks, and traces that in the space of the creature profanely a-wait the redemption—it is possible "die Geschichte gegen den Strich zu bürsten," to pass history "against the grain," overturn its form of empty duration (be it linear or cyclical). It is possible to discern how the radical incompletion of the world does not necessarily produce, demonically, mere desperation due to the breakage but permits, within the thing, to surprise the unexpected meantime of a *Yes* that is stronger than any fall and any consumption, of a *But* that snaps the infinite repetition of the catastrophe. Tiqqun, weak and continuously foundering,[12] frees not only from the brothels of historicism but also from the subtler fascination of the investigators of the future, and guards, beyond any mythology, the idea of prophecy: not vision of the future but salvation of *every* moment in its being able to *name* itself as that moment, that meantime in which the symbolic primacy of the word can represent itself, and precisely at the very peak of the allegorical, amid its ruins. Projected onto every event is the shadow of this eschatological "reserve," strong enough to free us from every "been" and chrono-latry.

TRANSLATED BY HOWARD RODGER MACLEAN

EMANUELE SEVERINO
Time and Alienation

"Happy the man who can say 'when,' 'before,' and 'after'!" So writes
Robert Musil in *The Man without Qualities*. In *De interpretatione* Aristotle con-
veys his own conception of bliss: "When what is, is, it necessarily is; and
when what is not, is not, it necessarily is not. But it is not of necessity
that everything that is, is; nor that everything that is not, is not. That
everything that is necessarily is, when it is, is not the same thing as being
purely and simply of necessity. The same must be said as regards what is
not" (19a23–27).[1]

"What-is" is *to on, being (ente)*: the participle *on* indicates not simply
"is" (*estin*), but, precisely, *what-is*, the synthesis of a certain determination
(e.g., house, star, man) and its *Being (essere)*. Accordingly, Aristotle's text
states first of all that *when (hotan)* what-is—say, a house—is, *then* indeed
the house necessarily is; but not that it necessarily is *haplōs, tout court*; i.e.,
the house does not exist of necessity. In fact, just as we say, "When a
house is," so we also say, "When a house is not." A house "is not" when
it has not yet been built, and when it has been destroyed. The phrase
"when a house is not" means either "before it was" or "after it has been."
Thus all the occasions of Musil's "bliss" are present.

But this bliss now dominates the earth. Greek thought established
once and for all the meaning of "when," of "before," and of "after," relating
them to *Being* and to *not-Being* (to *estin* and to *mē estin*). The whole of Western civilization grows within this rigorously consistent "bliss" (even when
we think Greek ontology no longer concerns us). And yet it is the very
symptom of alienation. In spite of everything, Western civilization still remains within the meaning that the Greeks gave to time. Indeed, time itself coincides with this meaning. But the aim of these pages is to recall,
once again, that *time* is the very essence of *alienation*. And the essence
of alienation is *essential* alienation, infinitely more radical and infinitely
deeper than any religious, economic, psychological, or existential alienation.

When a house is not, Aristotle says, it is *mē on*, non-being. Western
man is concerned with establishing that when what-is-not is not, it necessarily is not. But he leaves in the deepest and most unexplored darkness

the meaning of the expression "when a house is not" (or "when a man, trees, stars, the earth, love, peace, war are not"). That which, when it is not, is, is—for example—a house. When a house has been destroyed and has become something past, it is not. Normally one adds the word "longer" and says it "is no longer": but, indeed, that which is no longer is not (and that which is not yet is not). Thus it is *of a house* and *of men, of stars, of the earth, of love, of peace, of war*, that Western man says that they are past and therefore are not. But a house (and the other things said to "pass") is not a Nothing. A house is a place that shelters mortals from the harshness of the seasons; it is the openness of a determinate meaning: "house" does not signify "nothing" (the meaning in which a house consists does not signify "nothing"), and for this very reason a house is not a Nothing. In fact, Western man's very language draws a distinction between these two phrases: "when *a house* is not" and "when *a Nothing* (or Nothing) is not." Such language does not believe that it can replace the phrase "when a house is not" with the phrase "when a Nothing is not," precisely because it does not take to be a Nothing that which it affirms is not. Which means that Western thought affirms *of a not-Nothing* that when it is not, it is not: for this is affirmed *of* a house, *of* a man, *of* the earth, *of* what is past, *of* what is future. "When a house is not" therefore means "when a not-Nothing is not," which is to say, "when a being is not." In stating that it is not of necessity that all that is—*to on hapan*—is, Aristotle is indeed stating that some being can not-be: and it is *of this being* that, when it is not, one must say "when a being is not." To be a being means in fact to be a not-Nothing. Accordingly, in the Aristotelian affirmation "when what is not [*to mē on*], is not, it necessarily is not," the term *mē on* does not indicate Nothing and thus the affirmation does *not* mean "when Nothing is not it necessarily is not." In this affirmation the term *to mē on* does not indicate Nothing—on the contrary, it indicates *not*-Nothing or *being* (in its happening to not-be), and hence the phrase "when what is not is not" signifies "when a not-Nothing, that is, a being (e.g., a house) that happens to not-be is not." Or "when *being* that is not is not."

When a house, once destroyed, becomes something past (or when it is still something future), it is not. For Western man, what "passes" does not pass completely: something remains of what is past. Memories, traces, regrets, hates, consequences, effects—these remain. But not *all* of what is past remains. If all should remain, then *nothing* would be past. What remains, we say, "is"; what does not remain "is no longer." When a house is no longer, something of it does not remain. Ruins and memories remain, but something does not remain, something "is no longer." For Greek thought and for all of Western civilization, saying that something "does not remain" and "is no longer" means saying that it has become a Nothing. It is true that from Aristotle to Marx the destruction of a house is not its total annihilation (precisely because something remains even after its destruction)—yet it is also true that, for Western thought, with the destruction of a house at least *something* of the house must become a

Nothing. At least the unity and form that the materials of the house possessed, when the house was, become a Nothing; as does the irreplaceable atmosphere created by this unity and form. When a house is not, something of the house (at least the specific unity of the materials of which it was made) has become a Nothing. And it is precisely of this *something* become a Nothing that we are thinking when we say "when the house is not." So for a house to be destroyed and become something past and be no longer, at least something of the house must become a Nothing: if *nothing* of the house became a Nothing, Western man would not even say that the house has been destroyed, that it is past, that it is no longer. And this something that belongs to the house, once again, is not a Nothing but is *being*—it is the specific unity of the atmosphere and materials of the house that was destroyed. It is this *being* that, when a house becomes something past, becomes a Nothing. In saying "when a house is not," by the word "house" language does not refer to that which remains *of* the house (ruins, memories) but rather to that very *something*, to that being which becomes nothing and which is less a something that belongs *to* the house, that is part *of* the house, than it is the *house itself* as a specific and irreplaceable way and atmosphere of dwelling.

"When a house is not" means therefore: "when a being is nothing." The phrase "when the sky is blue" contains the affirmation "the sky is blue." And thus the phrase "when being is nothing" contains the affirmation "being is nothing." If one asks a Western man if being (a house, a man, a star, a tree, love, peace, war) is nothing, his reply will most certainly be *no*, being is not a Nothing. Yet for more than two thousand years he has continued to say of being "when being is not," and he goes on thinking that being is nothing. And he continues to experience being as if it were a Nothing. If someone were to say "when the sun is the moon" or "when the circle is a square," "when stones are birds" or "when even is odd," anyone in the West would immediately respond that a "when," a time in which the sun is the moon and the circle is a square, stones are birds and even is odd is not possible. But this feeling for the absurd does not prevent him from thinking "when being is nothing"; it does not prevent him from thinking *that* being is nothing. That bliss which Musil praised (but the whole Western world is of one mind in praising it) postulates the persuasion that being is nothing. "Before" means "before being is," and one can say "before being is" when being is not (yet), or when it is nothing; and "after" means "after being is," and one may say "after being is" when being is no longer, or when, once again, it is a Nothing. The persuasion that *time is* postulates the persuasion that *being is nothing*. Time may exist, in fact, only if a "when being is not"—a "when being is nothing"—exists; therefore time may exist only is being is nothing. The nothingness of being is *nihilism*, and nihilism is essential alienation. Western civilization grows within the persuasion that being is *in time* and thus is *nothing*.

All this seems, to the eyes of Western man, to be based upon pseudointellectual subtleties. He objects: "When being, by becoming something past, has become a Nothing, it is a Nothing. When it is nothing, it is nothing. Hence it is not true that, in positing a time when being is not, one thinks that being is nothing." When being is nothing, it is nothing, he says. But essential alienation consists precisely in the persuasion that there exists a "when" it—namely, *being!*—is nothing. Western man establishes an identity between the Nothing and the "when" (that is, the time in which) being is a Nothing. But this apparent identity between Nothing and Nothing conceals the identity between *being* and Nothing, namely, that identity which constitutes itself when one accepts *time*, the "when being is nothing."

It is believed that *tempus* and the *templum* alike are a *temnein*—a *separating* of the sacred from the profane. But *tempus* is a separation abysmally more radical than the separation of the sacred from the profane. *Tempus* separates beings from their Being, it separates the "what" from its "is": only on the ground of this original separation may one conceive a "when" being (the "what") is united *to* Being, and a "when" being is separated *from* Being (i.e., a "when being is" and a "when being is not"). Original separation of being from Being, as the essence of time, shows that being, as such, is a Nothing: in order to not-be a Nothing it must be united to that Being from which it was originally separated. In testifying to this fundamental meaning of *tempus* (the Greek *khronos* still echoes the word *krinein*, i.e., to separate) Greek thought brings to light the hidden, implicit ground upon which the separation of the sacred and the profane rests in archaic preontological (i.e., pre-Western) civilizations; whereas in Christianity the fundamental meaning of *tempus*—the Greek meaning of time—becomes the explicit ground of the separation of the sacred from the profane. It is because being—a stone, a tree, a star, a man's life, the earth—is originally separated from Being and is experienced in this separation, it is because being is experienced in *time* that it finds itself abandoned to nothingness and goes in search of a source, an *axis mundi*, a god, something sacred, or a kerygma that guarantees its union with Being. It is because man lives in time—which is to say, in essential alienation—that he builds *templa* and evokes the sacred, be it the cosmic sacred or the historical sacred of the Christian kerygma. But it is also because man lives in time that he entrusts his salvation to modern science and the technology to which it gave rise, when he realizes that the sacred cannot save him from nothingness. The sacred and technology are the two fundamental ways in which the inhabitant of time, that is, of essential alienation, seeks his own salvation, or seeks to save that being which is his world and his life, anchoring it to Being. But such salvation is impossible—for it does not transcend alienation but rather attempts to survive within it. Since inhabiting time means separating being from Being, to *will* this separation is to will the impossible (because *that being is not*—namely, that being is nothing—is the epitome of impossibility, and salvation is impossible precisely because it is the will to survive within impossibility); and yet it is this "will to the impossible" that, as scientific-technological will, now dominates the earth.

Paul Ricoeur, for one, has attempted a "mediation" between the cosmic sacred and the Christian kerygma, as opposed to the program of demythologization and desacralization of the Christian message and separation of religion and faith. The basis of this program is the recognition that science has destroyed the universe of myth. But, for Ricoeur, if science has eliminated the sacred from the modern world, the ideology of science and technology has now itself become a problem. And he finds allies for his thesis in Heidegger, Marcuse, Habermas, and Ellul. Referring to Habermas—for whom "the interest of empirical knowledge and the exploitation of nature is limited to that of practically and theoretically controlling the world of man," so that "modernity takes the form of the boundless extension of a single interest at the expense of all others, and above all at the expense of an interest in communication and liberation"—Ricoeur affirms that "modernity—namely, scientific-technological ideology—is neither a fact nor a destiny: it has become an open question."[2]

But if time is the essential alienation in which the existence of mortals grows, then scientific-technological domination of being and the consequent destruction of every mythical universe and of every kerygma not only are a fact but also are the destiny demanded by the essence of time. For the inhabitants of time, "modernity" is objectively a *closed* question, even if some among them cling to the illusion of being able to open it.

For the Time-dwellers, time is "original evidence." It is "evident" that the beings of the world are that of which it must be said "when it is" and "when it is not," or "when it is not a Nothing" and "when it is a Nothing." Being is that which issues from and returns to nothingness. When being had not yet issued from nothingness, it was a Nothing; when it returns there, it is a Nothing once again. But only because being is in time—only because being is thought and experienced as a Nothing—can the project of guiding its oscillation between Being and nothingness arise. Only on the basis of *time* is the *domination* of being possible. And, in the openness of time, the birth of the project of dominating and exploiting being is not only possible but inevitable. Inhabiting time is the very essence of this project. Time is in fact that separation (*temnein*) of being from Being which takes possession of being as that which can be assigned to Being (from which it was originally separated) and to Nothing, confronting it in its availability to the decision that so assigns it. With this separation being becomes an absolute availability to the forces that tear it away from and thrust it back into nothingness. The will that being be time—the will that wills that the meaning of being be time—is the original form of the *will to power*. The original will to power, which takes possession of being separating it from Being and making it available to domination, is precisely the will to guide and control being's oscillation between Being and nothingness.

The will that drives its domination of being to the point of identifying it with Nothing—driving it to the remotest distance from itself—is the original project *destined* to be realized as the scientific-technological domination of being that destroys the domination of being attempted through sacralization of being, religious invocation, and Christian faith. In fact, the will to power first dominates being by conjoining it with the sacred and the archetype, namely, with the source of Being.

Mircea Eliade, who is a point of reference for Ricoeur, recognizes that if archaic languages lack such terms as "Being," "non-Being," and "Becoming," the *thing* (the fact) of "Being," "non-Being," and "Becoming" is nevertheless present. And the "thing" is that beings (both human and nonhuman) become sacred only insofar as they participate in the Being of an archetypal world that transcends them. Removed from this participation, beings become "the profane world" that, as Eliade says at the beginning of *The Myth of the Eternal Return*, "is the unreal par excellence, the non-created, the non-existent: the Nothing."[3] The will to power dominates being by immersing it in the sacred, that is, in Being. But Eliade maintains that for archaic man the immersion of being in the sacred is cyclical and that this cyclical return of being to the sacred "betrays an ontology uncontaminated by time and Becoming." Nevertheless, the return to the sacred—the will to be as the archetypes are—is, for Eliade, the way in which archaic man "opposes," "endures," and "defends himself" against history. It is, in short, his way of dominating "history." But "history" is time. Precisely because Eliade's archaic man accepts time—and so lives in essential alienation—he attempts to defend himself against time and to master it through identification with the archetype. It is because he is an inhabitant of time that he attempts to master time both by fashioning an ontology not dominated by time and Becoming, and by restoring being to the original world of the sacred. The same thing occurs in Christianity and in all the formulations of Greco-Christian theology. The opposition affirmed by Eliade between archaic man's antihistoricism and Christian man's historicism remains within the acceptance of time. It is because mortals have separated, implicitly or explicitly, being from Being that man has need of God (or of revolutionary praxis, or of technology)—that is, of a *ground* of being. Jesus wants to save man and give him eternal life because Jesus too is an inhabitant of time and sees around him only beings abandoned to nothingness and so in need of salvation. The search for salvation (which is one and the same with the project of dominating being) is an expression of the essential alienation of man. When men such as Rudolf Bultmann or Dietrich Bonhoeffer demythologize the Christian message and separate faith from religion, they too inevitably remain within this alienation. Their endeavor is based on the consciousness that the sacred is powerless to dominate being, and that salvation (domination of being) must be pursued in some other way.

For success, power, and domination and exploitation of being is the destiny of whoever dwells in time. To dwell in time is to dominate and domination demands the destruction of every form of domination that proves to be powerless. Science and modern technology have shown the impotence of the domination of being through union with the sacred and with God. The power of technology has shown the impotence of the sacred and of God, just as it has shown the impotence of every ideology that, like Marxism, purports to dominate the earth. For the Time-dwellers, "modernity"—scientific-technological power—is the destiny of the West. The openness of time is the original power, and the logic of power requires that every power fall before a power more powerful. That the interest constituted by theoretically and practically controlling man's environment should expand at the expense of all less powerful interests, such as those of communication and liberation (this is Jürgen Habermas's critique, taken up by Ricoeur), this de facto encroachment of the will to mastery is itself the irrefutable *reason* why the interest constituted by scientific-technological will to mastery is destined to destroy all other interests. To inhabit time is to inhabit the logic of power, and this logic decrees that the force which is *in fact* more powerful is destined to dominate all other forces and alternative interests. Spirit, human dignity, values, brotherhood, love, liberation, morals, politics, the sacred, God, Christ—all the forms of Western civilization matured within the acceptance of time—have progressively proved to be powerless when confronted by the power of technology. They have proved to be impotent forms of the will to power. Their destruction therefore is not only a fact to be recorded but also the destiny that can no longer be avoided since mortals dwell in time.

The triumph of technology is the triumph of nihilism. Much of contemporary culture recognizes this fact. But Western culture has not recognized the essential *meaning* of nihilism. Ricoeur affirms that "both the scientistic illusion and the retreat of the sacred . . . derive from the same forgetfulness of our roots. In two different but converging ways, *the desert grows*. What we are on the point of discovering, in spite of scientific-technological ideology, which is also military-industrial ideology, is that man is absolutely not possible with the sacred . . . man must not die."[4] But why must man be possible? Why must man not die? It is clear that Ricoeur is speaking of man in terms of value; but why must this value not die? Since time is the *meaning* of being, the essence of being is its ability to be destroyed and constructed, created and annihilated. Since being is availability to Being and nothingness, being (and thus also man) is destined to be manipulated, violated, and exploited by gods, masters, and technologies (as B. F. Skinner's *Beyond Freedom and Dignity* attests). Technological power and the destruction of the sacred and of the kerygma do not imply forgetfulness of our roots, because our roots are our dwelling in time, and science and technology are the most rigorously consistent realization of this dwelling. To be sure, the desert grows. But the desert is time, and the destiny of this growth is scientific-technological domination of the earth. All the Time-dwellers who—with Heidegger, Adorno,

Marcuse, Habermas, Fromm, Ellul, Ricoeur, and many others—seek to oppose the desert's growth and to defend man and his dignity, all those who belong to the culture that condemns technological civilization, inevitably fail because they are not true to their authentic roots (that is, to essential alienation), because they are not consistent with the essential persuasion that envelops them. Their aspirations and projects of a more human world are the wreckage that the desert's relentless growth leaves behind. Philosophy, Christianity, Marxism, art, are the colossal wrecks of this ever-growing desert.

Just as one cannot combat a disease by restoring the physiological conditions that originally caused it, so one cannot oppose the desert's growth by returning to traditional or archaic forms of human civilization. Essential alienation appears only insofar as does truth, with respect to which alienation is and shows itself as such. Parmenides, the most misunderstood thinker in the history of man, took the first step in testifying to truth when he said, "You shall not sever Being from its connection with Being," *ou gar apotmēzei to eon tou eontos exesthai* (frag. 4). But Parmenides' testimony remains a presentiment. The gaze that sees the desert growing and that sees its authentic meaning does not belong to the desert.[5] In this gaze being, all being, every being, from the most humble to the most solemn and exalted, is originally united to Being. In this gaze *all* things share the nature of the sun, whose existence continues to shine even when nightfall hides it from our eyes. The Time-dwellers created both the gods and the jealousy of the gods: the gods are jealous because they kept for themselves that unity with Being which is the property of *every* thing. In this gaze every being is eternal (*aiōn*, i.e., *aei on*, or united immediately to *estin*), and the variation of the world's spectacle, the appearing of variation, is the rising and setting, the showing and the hiding of *the eternal*, in every way like the sun.

But this opens up another aspect of the question, developed elsewhere: the discourse of the hermeneutics of the appearing of Being.[6] Here I can merely indicate its general course by saying that the persuasion that time is evident—the persuasion that time *appears*, that the separation of being from Being *appears*—belongs itself to that essential alienation which is what time *is*. The Time-dwellers believe that time, the "when things are not," is *visible, manifest*. But for them it is unquestionable that when a being—say, a house—is not, not only has it become a Nothing, but it also ceases to *appear*: when a house has been destroyed and is no longer, it no longers *appears* either, to the extent that it no longer is. But this means that *on the basis of Appearing we cannot know anything* about that which, "having become a Nothing," no longer appears. In "being destroyed" and "becoming a Nothing," the house leaves Appearing; and Appearing, as such, shows and says nothing about what befalls a being that has left its horizon. Therefore having been a Nothing (when the house is not yet) and becoming a Nothing anew (when the house is no longer) cannot appear; the nothingness of being—namely, *time*—is not something that *appears*, it is not itself a "phenomenological" content. The persuasion that time appears is therefore the result of a hermeneutics of Appearing that, on the ground of the will that being be nothing, wills that the nothingness of being be something visible, manifest, and evident.

Outside of essential alienation, that which appears is being—the immutable, the eternal. The eternal enters and leaves Appearing, just as the sun—which shines eternal—enters and leaves the vault of the sky. When being leaves the vault of what appears, Appearing keeps silent about the fate of the being that is hidden (and "when" assumes an unheard-of meaning). But the Erinyes of truth (*Dikēs epikouroi*) of whom Heraclitus speaks (frag. 94) reach what is hidden and remind it of its destiny: the Necessity, the *Anagkē* that it remain united to its Being.

TRANSLATED BY GIACOMO DONIS

EMANUELE SEVERINO
The Earth and the Essence of Man

1. The Body and Being as Tekhnē

"How, with the death of a man, is the soul not dispersed and this not the end of its Being?" ("Hopōs mē hom' apothnēskontos tou anthrōpou diaskedannuetai hē psykhē kai autēi tou einai touto telos hēi," *Phaedo* 77b). This is essentially a *metaphysical* problem. Not because it considers the relationship between the "here" and the "beyond," but because it admits the possibility of the "end of Being" (*telos tou einai*), that is, its annihilation. It is a specific way of asking whether a certain being continues to exist even when a certain other being exists no longer. Thus the fundamental presupposition is that beings *can* not-exist, and so also can exist no longer, that is, can "end."

For metaphysics, things "are." Their "Being" is their not-being-a-Nothing. Insofar as they are, they are said to be "beings" (*enti*) or "Beings" (*esseri*). But being, *as such*, is that which *can* not-be: both in the sense that it could not-have-been or could not-be, and in the sense that it begins and ends (was not and is no longer). Metaphysics is the assenting to the not-Being of being. In affirming that being is not—in assenting to its nonexistence—metaphysics affirms that the not-Nothing is nothing. Precisely because the fundamental notion of metaphysics is that being, *as such*, is nothing, metaphysics must seek *reasons* to support its thesis that certain privileged ("divine") beings are exempt from birth and death, and so cannot be said to not-be. These "reasons" alone enable it to recognize the essential not-nothingness of certain beings; without them, being as such appears to it as a Nothing.

Today we no longer believe in the metaphysical reasons for the immortality of the soul. And yet, with the latest developments of science, the project of practically constructing precisely that which metaphysics was unable to demonstrate grows ever more determinate and consistent. But this project too—like the entire history of the West—grows within the fundamental notion of metaphysics. For the construction of being can be undertaken only if being is thought as what can begin and end; or, in general, as what can not-be.

Western culture is incapable of setting any limit to technology's aggression against being. The project of constructing man's body is now inseparably accompanied by the project of constructing mental facts. Thus human happiness is no longer seen and pursued as a transcendent condition, determined by man's moral conduct during his life in the world (or by the combination of such conduct and divine grace), nor as an immanent result of historical dialectic. Happiness is seen today as the product of a technology whose success derives from its being rooted in physico-mathematical knowledge. Western culture can set no limit to this aggression against being, because the *essence* of such culture is metaphysical nihilism, whose most radical and consistent realization is technology itself.

From the very dawn of metaphysical thought, Being has been *tekhnē*. In the *Sophist* (247d-e) Plato defines Being as *dunamis* (power): that which is (*to on*) is that which has the power of making or of being made: *dunamin eit' eis to poiein eit' eis to pathein*. "Making" (*poiein*) signifies bringing into Being (*eis ousian*) that which previously was not (*hoper an mē proteron on*); "being made" (*poieisthai*) signifies being brought into Being (*Sophist* 219b). But power is the very essence of *tekhnē*, because if *tekhnē* can be divided into productive *tekhnē* and acquisitive *tekhnē* (*poiētikē tekhnē, ktētikē tekhnē*), the acquisition of beings—such as money making, property holding, hunting, fighting, knowledge—is nothing but an ordering of that which has already been produced in the various forms of *poiētikē tekhnē* (219c). The distinction between divine *tekhnē* and human *tekhnē* (*theia tekhnē, anthrōpinē tekhnē*, 265b–e) is therefore the supreme difference between beings. *Theia tekhnē* produces all the beings of nature; *anthrōpinē tekhnē* produces all the beings which are brought from not-Being to Being in human arts. Being is *tekhnē* because it is essentially enveloped by the horizon of making and of being made; that is, because it belongs essentially to the process of bringing and of being brought from not-Being to Being (*aitia tois mē proteron ousin huteron qiqnesthai*). If something is not *tekhnikon*—if it does not produce or is not produced, or is not part of the process of producing–being produced—then it *is not*: it is a Nothing. *Theia tekhnē* has today been supplanted by *anthrōpinē tekhnē*, but the meaning of Being has remained identical to the one established by Plato once and for all in Western history. God and modern technology are the two fundamental expressions of metaphysical nihilism.

2. The Eternity of the Body and the Spectacle of Alienation

Authentic untimeliness[1] is the overcoming of the essence of the West. But, above all, it testifies to the truth of Being, which says that Being is and that it is not possible that it not-be (*hē men hopōs estin te kai hōs ouk esti mē einai*, Parmenides, frag. 2, v. 3). "Being" means everything that is not a Nothing. But only *the Nothing is* nothing. "Nothing" cannot also be predicated of a "something," which is presumed to be meaningful as a not-Nothing (and any meaning whatsoever is meaningful as a not-Nothing), and is, at the same time, relegated to the limbo of nonexistence—for it is posited as "something" (namely, a not-Nothing) that, when it is not, is nothing. Thus thought that testifies to the truth of Being cannot accept the claim that with the death of the body the soul continues to exist—not because it claims that, when the body no longer exists, the soul cannot exist either, but rather because both body and soul are *eternal*. Like *every* being. The soul cannot exist without the body, just as it cannot exist without *any* being, for the destiny of *all* being is to exist. ("Aeternus" is a syncope of "aeviternus," and "aevum" is *aiōn*, "always being," the impossibility of not being. But this impossibility must be referred to the totality of Being, not to a privileged being—and therefore the Greek *aiōn* is the very expression of metaphysical nihilism.)

The body's disintegration is not its annihilation but rather is the way in which it stably leaves the horizon of the appearing of Being. History is the process of the appearing and disappearing of the eternal. Dialectic is not the essence of Being insofar as it is, but of Being insofar as it appears. *Being* cannot be altered by the onslaught of technology. Unscathed, it uncovers the spectacle of the alienation of the meaning of Being—the spectacle of our time. Man today believes he can attain unlimited control of the creation and annihilation of Being. This persuasion is the basis of every work he performs, which means that every work brings into Appearing the spectacle of alienation. If we were convinced that by opening and closing our eyes we caused the birth and the annihilation of visible things, we could certainly develop a way of living based on this conviction; but the reality that would appear and the life we would live would be different from the reality that would appear and the life we would live if we were free from this form of alienation. Like this movement of the eyes, Western technology too is an art of disclosing Being. But this art brings into Appearing a different content from what would appear if the West were free from the alienation in which metaphysical nihilism consists. The technological construction of man does not invent man but is the *disclosing* of *eternal* man. Technology, however, in its failure to see that its—and all—acting is essentially a revealing, discloses a different humanity from that which would appear in the light of the truth of Being. It discloses the humanity of alienation.

3. The Coherence of Technology

Everything is eternal. And so, also the *appearing* of Being is eternal. But while in Appearing there are things that appear and disappear, Appearing itself, as the total horizon, cannot appear and disappear. If it appears and disappears, then it is only the appearing of a part of what appears, while Appearing, as the transcendental event, is the locus in which everything (and so also the appearing of certain things) begins and ceases to appear. But in the truth of Being it cannot even be *supposed* that everything has ceased to appear (or might never have appeared): if that should occur, Appearing (i.e., a not-Nothing) would become a Nothing. Being is *destined* to appear. In this destination lies the essence of man. The original meaning of 'soul'—of 'mind', 'thought', 'consciousness', etc.—is its positing itself as the appearing of Being. And 'I' signifies Appearing insofar as it has itself as its content; that is, it expresses in condensed form the identity of form and content.

Not only is man eternal, like every being, but he is also the locus in which the eternal eternally manifests itself. The metaphysical alienation of the West is inevitably accompanied by an inability to comprehend the meaning of man. Appearing is understood either as an empirical determination (a being among the beings that appear) or as the transcendental horizon. In the first case, man too is a being that issues from and returns to nothingness. His consciousness is conditioned by birth and death; during his life it is continually being kindled and extinguished, as sleep and the phenomena of "loss of consciousness" traverse it. Thus technology can undertake the construction of a consciousness free from these conditioning factors. For idealism, on the other hand, consciousness is the transcendental horizon, containing time within itself; and therefore it is eternal. But in this case, eternity expresses the ontological privilege of thought with respect to beings that are thought, just as Plato's Idea is privileged with respect to sensible beings. Thus, like every metaphysical demonstration of the existence of an immutable being, idealism's grounding of the eternity of thought is destined to fail. And technological projecting of man can therefore legitimately undertake the construction not only of particular mental facts, or of consciousness understood as one of the particular facts of experience, but also of thought's transcendental horizon itself and its incorruptibility.

Any philosophical-metaphysical protest raised by Western culture against these alleged excesses of technology overlooks the fact that technology takes the fundamental thought upon which both the protest and all Western history rest to its logical conclusion. In undertaking to transform man into superman and God, technology nonetheless operates within the horizon that, opened up for the first time by metaphysics, encloses the entire development of our culture. Technology takes to its logical conclusion the meaning of the metaphysical horizon by which it too is enveloped—the horizon constituted by the thought that being can be a Nothing. This thought has been the basis of every metaphysical affirmation of immutable being. It is therefore perfectly consistent that metaphysics *qua* technology should undertake the practical construction of the immutables and the immortals that metaphysics *qua* contemplation has been unable to ground. Except that, in so doing, metaphysics no longer brings to perdition mere modes of reasoning but the entire civilization of men on earth. Metaphysics *qua* technology has in fact transformed everything that appears—the customs of peoples, houses, plants, the stars— into a spectacle of perdition.

4. The Never-setting and Philosophy

Within the horizon of all that appears, the great stream of determinations that appear and disappear is held in by never-setting banks: they accompany those beings whose Appearing is necessarily implied by the appearing of any being. These beings are the never-setting "background" of any disclosure of Being, the eternal spectacle in which all time—and so also the history of the alienation of the West—unfolds.

The appearing of a being is necessarily implied by the appearing of another being, when the first being is a necessary determination of the second. Originally, the necessary determination of Being (and so also of Being that appears) is the truth of Being. Truth is the incontrovertible position (positing) of its content. It says of the content what it is Necessity to say—and is the original openness of the sense of 'necessity'. Truth is the incontrovertible appearing of the totality of Being, insofar as that totality is dominated by the necessity that opposes Being to not-Being. Since truth is the structure of the necessary determinations of what can be affirmed with truth, no Being can appear without the appearing of the truth of Being. A being—this book, for example—is not its other (is not other than what it is): being the negation of its other is a necessary determination of this being. But this can occur only if the possibility of calling into question the position of such being and of its predicate has been superseded, and only if the incontrovertible meaning of necessity is manifest. A necessary connection is such only insofar as it is inscribed in the original structure of the truth of Being, which accordingly—as the structure of the necessary determination of Being—is the never-setting background that accompanies and envelops any manifestation of a Being.

Philosophy does not guard a truth that man happened upon at a certain moment in his history: man is the eternal appearing of the truth of Being. And philosophy is the emerging of this essential hearing—which is the very essence of man—once every other hearing has been relinquished. Philosophy does not present us with new things, previously unknown, but is the conspicuousness that what has always stood before us assumes when attention is no longer focused on what supervenes and vanishes. Not only are we eternal, but—since the eternity of Appearing belongs to the truth of Being—we eternally *know* we are eternal. Only within this immutable appearing of the truth of Being can the history of man—and so also the history of the West, as the history of the abandonment of that truth—unfold. Distraction from truth is possible only insofar as truth continues to appear. For men, "oblivion of the sacred spectacle" (*lēthē hōn tot' eidon hierōn*, *Phaedrus* 250a) is a concealment of truth, in the sense in which it can be said that the sky conceals itself from a countryman watching birds in flight or falling stars: the sky of truth appears eternally, but the things that cross it call attention to themselves and become all-important. Then, language has words only for what is important and life runs its course, as if truth had set and only things remained.

5. The Occurrence of the Earth

The beings crossing the never-setting sky of the truth of Being are the occurrence. The background does not occur but is the still place that receives the occurrence. Being eternally appears in its truth, and in this Appearing, the flowering of Being occurs. In the clear silence of truth the occurrence is the prodigious. However long awaited, it is in fact the unexpected. Since it receives the occurrence, Appearing is not the infinite appearing of Being, the epiphany in which the completed totality is disclosed and in which, therefore, no further revelation can occur. As finite Appearing, eternal truth is contradiction. Although it lets all things appear in the whole, the whole in which it envelops them is not the completed totality of Being but only the formal meaning of this totality. Thus what is not the whole is made to appear as "whole." This contradiction could be superseded only if finite Appearing should become infinite. The eternal appearing of the background is the manifesting of a contradiction, which could be resolved only in an occurrence. But truth also knows that the whole cannot become an occurrence.[2] Thus, in its essence, truth awaits the occurrence granted it: the measure of its liberation from contradiction. In this measure lies the salvation of truth (i.e., the truth of salvation). However long awaited and however much truth knows of it, the occurrence is the unexpected. If truth knew everything of it, the occurring of the occurrence could add nothing to such knowing, within which the ocurrence would therefore have always already occurred.

The prodigious occurrence is the earth. Joy and pain, war and peace, feelings, stars, thoughts and actions all belong to the earth. The earth is Being's offering. Being has always inhabited Appearing, but the earth is the guest's long-awaited gift. Truth accepted the offering. In the beginning, truth willed the earth, and this will encloses and sustains any mortal willing.

In the life of man, philosophy is an unusual event. Man normally lives in untruth, looking after the problems of the earth—the problems, that is, of everyday life and those raised by religions and ideologies, by science and art, and by philosophies themselves. But the life of man is, in its essence, the eternal appearing of Being; and Being can *only* appear in its truth, since the truth of Being is the background whose Appearing is necessarily implied by the appearing of any thing. However deep the untruth in which he lives may be, man is still the eternal manifestation of the truth of Being. Living in untruth cannot, therefore, be thought to be an oblivion that leads to the disappearing of the truth of Being. For untruth is possible only within that truth: not insofar as it is "a part" of truth, but insofar as it belongs to the occurrence that comes to light in the eternal appearing of the truth of Being.

Solicitude for the earth, in which untruth consists, is grounded first of all on the truth of Being's acceptance of the earth. We can will something—a house, food, love—only insofar as we first will the horizon within which the individual things that we will can appear. The occurrence of the earth is the originally willed horizon, in which any thing that we will is willed. But this original will would not be possible if the eternal appearing of truth had not accepted from Being the offering of the earth.

This receiving of the earth, performed by the truth of Being, is the same original that acts in the solicitude felt by untruth for the earth. But in untruth, the receiving of the earth unites with the conviction that the earth is the whole with which, assuredly, we deal. In this conviction, Being that occurs is *isolated* from the truth of Being. Untruth is possible only insofar as the occurrence brings with it, in the eternal appearing of the truth of Being, both the receiving of the earth and the isolating conviction. For receiving the offering belongs to the offering: it is the way in which the occurrence occurs. But also the isolating conviction belongs to the offering. For, if the background is indeed the *truth* of Being, error (the isolating conviction) can (and must) belong *as negated* to the background; so that, *as posited*—as that of which one is convinced—it cannot but occur with the occurrence of the offering.

As willed by truth, the earth stands out against the background. Truth, in its receiving of the earth, wills that the earth continue to appear. Receiving a guest means willing that he remain. In willing the continuation of its occurring, truth does not treat the occurrence as the unexpected. The occurrence, as such, is the unexpected, but *willing* the occurrence means no longer treating the unexpected as unexpected; it means giving the unexpected what it does not have. This giving—namely, the will that the occurrence continue to occur—is the conspicuousness of the earth. To the *earth*, as projected into the future (into the place where it continues to occur), is given what it does not have.

But however much the earth may stand out against the background, the receiving of the earth cannot conceal the background—cannot, that is, conceal the truth of Being. The earth is received in the light of truth: in the eternal appearing of the truth of Being, Being flowers, this flowering is the earth, and the receiving of the earth is the way in which the flowering is spread out in Appearing. But for that kind of distraction from truth to arise, in which untruth consists as the normal condition of the life of man, something else is required besides the receiving of the earth. Untruth is solicitude for the earth, united with the conviction that the earth is the dimension with which, assuredly, we deal, and beyond which there is total darkness. Since the appearing of the truth of Being is eternal and never-setting, that other which (for untruth to arise) is required besides the receiving of the earth cannot be the setting—that is, the disappearing—of truth, but must be the very conviction that the earth is what surely appears. In other words, the *other* is the appearing (i.e., the occurring) of this conviction within the never-setting appearing of truth. For Hegel, like Plato before him, truth (namely, that which from the viewpoint of metaphysical alienation is the truth of Being) appears only in philosophic consciousness. In other forms of consciousness (the forms of untruth), either truth is wholly absent, or it appears in a process, determining the dialectic transition to higher forms of consciousness. The myth of the cave corresponds to the phenomenology of spirit: truth, as a unitary totality, appears only at the end of a process. But the truth of Being neither rises nor sets, and in its eternal Appearing lies the essence of man.

Thus man, insofar as he lives in untruth, is the appearing of a *contention*: between truth, which eternally appears, and error, which accompanies the occurrence of the earth and sees in the earth the sure ground. In the appearing of the truth of Being, Being flowers and error belongs to its flowering: it is one of the beings that begin to appear. But it appears as at once denied and affirmed, rejected and accepted. Appearing, which as appearing of truth is negation of error, at the same time lets error stand free in Appearing as not negated, and therefore as accepted. In so doing, it becomes the scene of a contention: the appearing of a contradiction. Truth, which as such is already contradiction (because it posits as "whole" that which is not the whole: because it is the finite appearing of the infinite), here finds itself involved in a broader contradiction in which truth and error contend for Appearing.

But error's freedom in Appearing—its eluding the dominance of truth—remains an enigma. Is the appearing of the truth of Being itself responsible for this freedom, or is the rebellion against truth part of Appearing's destiny? Can error be "freely willed" by Appearing, or is the toleration of error—and so the existence of untruth—established by the necessity of Being?

[In response to the editor's request to shorten this chapter, the author selected the sections to be included (1-5, 10-14). The four central sections were dedicated to the discussion of more specific issues and can be omitted without essentially interrupting the discourse.]

10. *Threefold Alienation*

Let us review the fundamental traits in which the essence of man is revealed.

Being is eternal, and it eternally appears in this actual Appearing—which is not "mine" but which I myself am. Man has always been and will always be the revelation of Being, a satellite that forever accompanies the constellation of Being. Since the actual Appearing cannot not-exist (for it is itself a Being), Being is destined to appear—and therefore to appear in its truth, because the appearing of the truth of Being is that without which no Being can appear: it is the never-setting background of anything that appears. Being eternally appears linked to its "is" by dominant necessity; accordingly, the veritable and concrete meaning of necessity's dominance—of the structure of the truth of Being—eternally appears. As the eternal revelation of the truth of Being, man lives, in this sense, "the life of the gods" (*theōn bios, Phaedrus*). But the "plain of truth" (*to alētheias pedion*) stands gathered and still before him, and in this still spectacle man dwells forever. Contemplation is not a *periodos* outside of which man has a home to which he can return (*oikad' ēlthen*): his home is the truth that eternally stands before him.

Yet his original dwelling place is an infinite unrest. Any being that appears, appears included in the totality of Being, but this totality only appears formally: the concrete fullness of Being remains concealed. The eternal appearing of the truth of Being is the finite appearing of the infinite, where what is not the whole (because it is only the formal meaning of the whole—it is only this meaning, 'whole', without being the concrete to which this meaning is referred) is made to appear as the whole. In the eternal appearing of the truth of Being, the totality and every determination of Being appear as contradiction. The appearing of the truth of Being is the original being in contradiction. Being has always shown itself in Appearing, presenting itself in its truth. But Being with all its determinations does not enter Appearing. The occurrence of the earth testifies to the finitude of the primitive appearing of Being (everything that occurs is what has not yet appeared). In this primitive Appearing, therefore, the seal guaranteeing that all has definitively appeared cannot manifest itself. This means that the truth of Being, which eternally appears, includes finitude, namely, the essential contradiction of its own Appearing. This contradiction is the constitutive alienation of the essence of man. Supersession of this contradiction is the *absurd*: finite Appearing that becomes what it *cannot* become, namely, the infinite appearing of Being (see "The Path of Day," 17).

If this first alienation forms the essence of man, the occurrence of the earth beings with it a second and a third form of alienation. The second form is the occurring of the earth's isolation. Its conflict with the truth of the earth opens the horizon of man's living in untruth. This conflict gives rise to a second sense of being in contradiction, namely, the contradiction between the constitutive contradiction (in which the appearing of the truth of Being consists) and the earth's isolation. This second contradiction is man's life in untruth, his fallen existence. The third form of alienation is metaphysical alienation—namely, the history of the West. Here, the most gigantic effort is made to testify to the truth of Being. But the basis of this testimony remains the earth's isolation, which is just what in the authentic—and still unattempted—testimony should have been left behind as past. Greek metaphysics addressed itself to the truth of Being but did so without relinquishing the conviction that the earth is the region with which, assuredly, we deal. If the earth is the sure region, then the becoming of beings and the very occurrence of the earth have to be thought first of all as the process in which being has been, and returns to being, nothing. Metaphysics inquires into the conditions of the thinkability of Becoming so understood, which is to say, into the thinkability of the unthinkable. The history of the West has thus become a celebration of the solitude of the earth, and the West's gods are the gods of this solitude.

There is a fundamental difference, however, between the first indicated form of alienation and the other two. This is due less to our not knowing whether the other two forms also belong to the essence of man than it is to contradiction's meaning something different in them than it does in the first form. The first form is a contradiction not on account of what appears, but on account of what does *not* appear in it; not on account of what is said, but on account of what is *not* said. What is in fact said there—what appears—is only a part of Being, not the whole. Accordingly, the supersession (which moreover cannot occur) of such alienation would be the appearing of the whole: it would mean saying concretely *that very thing* which in alienation is said abstractly—and which *for this reason* appears as contradiction. In contrast, in the other two forms contradiction takes shape on account of what *is* said. In the second, the isolation of the earth is the negation of the truth of the earth, and this negation is held fast together with that truth. Here, there is not a not-saying but rather a saying no to truth. Just as in the metaphysical alienation of the West there is not a not-saying but a saying and a doing in the light of the thought that posits Being as identical to Nothing. One does not rid oneself of these other two forms of alienation by positing concretely that which, in them, is posited abstractly, but rather by negating it. In passing beyond the first form, it is the position of the content *qua* abstract that must become something past; in passing beyond the other two, what must become something past is the position of the content as such.

But metaphysical alienation has now become the dominant trait of the earth's isolation. The works of isolation have been overwhelmed by the works of metaphysics. Western civilization has become the supreme concreteness of the way in which the truth of Being is contested, and the occurrence of the earth made an object of contention between isolation and truth.

11. *The Earth's Isolation and the Mortal*
The isolation of the earth—which dominates the decisive moment of Western thought: the Platonic "parricide"—led metaphysics to think Being (determinations) as identical to Nothing; and it is upon metaphysical nihilism that Western civilization has been built. The civilization which today leads the peoples of the earth is grounded on the very event that brought about their fall into untruth. The earth's peoples have always been the dwelling places where the truth of Being gathers. But when the isolation of the earth came into these dwelling places, they became the houses of untruth. Which peoples have fallen into untruth?

Untruth is traceable first of all in the way in which the actual Appearing lets Being appear. I live in untruth. Which means: in the actual appearing of Being, the isolation of the earth continually counters the truth of Being. Only at times does the earth's solitude begin to set and let me begin to remain what I eternally am. The solitude soon returns, in full force and with all its consequences, so that all my decisions and works become decisions and works of untruth.

But the earth is always before me laden with the fruits of metaphysical alienation which were called out into the light by the people of the

West. The works which appear on the earth can in fact be interpreted as Being's response to the calls of the earth's peoples. Today, however, the works of the people of the West—the products of technological civilization—have overwhelmed all other works. Yet in *any* work—including those of the West—there is a trace of the truth of Being. In the works of untruth, there is also a trace of the earth's isolation. But we still have to learn how to uncover these traces. We know that any work can preserve the opposed traces of truth and of isolation, but what *are* these traces? What are the traces of truth? And what are the traces of the earth's isolation?

Metaphysical nihilism, within which Western history unfolds, is the trace of the solitude in which Western man has enveloped the earth. We can realize that the people to whom we belong have fallen into solitude because we know that metaphysics is the dominant spirit of the West. For metaphysics, being, as such, is nothing. This is the sign that metaphysical thought grows on the solitude of the earth. Isolated from its truth, the earth in fact is a Nothing. Isolation, which posits the earth as the surest region of being, is in its truth the nihilation of the earth. Indeed, affirming that the earth is the sure thing means, in truth, affirming that the earth is nothing. Metaphysics is the truth of isolation: it is the testimony to the nothingness of the earth. It betrays the truth of Being precisely because it looks at that truth through the solitude of the earth. Accordingly, it posits the totality of Being, just as the earth itself is posited in isolation. The isolation of earthly things thus becomes the transcendental determination according to which metaphysics posits the totality of beings. As isolated from their truth, all beings are a Nothing; and it is *precisely because* from the very beginning metaphysics thought being as a Nothing that it can explicitly affirm that being, as such, can become a Nothing. (For being as such is not incorruptible and ungenerable, but being insofar as it is a privileged being— insofar as it is one of the gods of the West.)

Isolated from its truth the earth is a Nothing, because if the earth can appear only insofar as its truth appears, then the appearing of the earth (or of any being whatsoever) without the appearing of that being's truth is what cannot be and therefore is a Nothing. Untruth is the fallen existence of mankind. It strikes root in the conviction that the earth is the sure region of being: the earth is the sure thing, because *the earth* is what appears. For the things of the earth to appear, nothing else is required but their Appearing itself. What a house, a man, a tree, joy or suffering is, is told by the thing itself in its unfolding and interweaving with the other earthly things that appear—for also in untruth beings are affirmed because they appear. Why do we affirm that the sky is blue, that we heard a voice, that the lamp is on the table (and so forth with the countless affirmations that make up the world in which Being appears in untruth), if not because the blue of the sky, the voice, and the lamp on the table *appear*, or we believe that they appear.[3] Even if only the content of Appearing and not Appearing as such is testified to, also in untruth the content is affirmed because it appears.

But untruth does not limit itself to affirming the earth. It also posits the earth as the thing that surely is and so sees in it the totality of the content that appears. In this way the earth is isolated from its truth (i.e., from the background that eternally appears and without which nothing can appear). Thus thinking that the earth is the sure region of being means thinking, in truth, that the earth is a Nothing, because—when referred to the earth—being the sure region means being a Nothing. For the earth is only a part (the part that occurred) of what truly is the sure region, and the part, when posited as the total content of Appearing, is a Nothing. (Furthermore: the part, thought without that on account of which it is—i.e., without its truth—is a Nothing.) In untruth, what is thought and therefore willed is the earth's nothingness: the things of the earth are treated as a Nothing in the very act in which they are posited as the sure region of being. Thus in any work of untruth a trace of the nihilation of things can be uncovered: every work bears the sign of the conviction that it is a Nothing. And so when language, as a work of untruth, intends to name the things of an earth left in solitude, its every word names the Nothing. But, that the earth should be in uncontested solitude is only an intention: the truth of the earth is the contrasting background that eternally appears, coming to light in every work and in every word of untruth.

Alienation, which makes man become a mortal, is the root of metaphysics. When he posits the earth as the sure region of being, man becomes a mortal—he becomes, that is, one of the things of the earth, whose nothingness is thought and willed. Metaphysics is the testimony to the nothingness of the earth. Isolated from the truth of Being, the earth stands before man as a Nothing. Metaphysics testifies to what stands before us and affirms that being, as such, is a Nothing. But metaphysics testifies to solitude not because it knows that isolation is the fall into untruth, but rather because it knows how to express the result of the fall. It does not express the fall as a fall (for this comes about in the truth of the untruth of Being), but rather as that which, as a result of the fall, lies before us (and before us lies the earth's nothingness). Thus metaphysics is not simply a false thinking—it is the consciousness that man *must* have of the meaning of Being and of himself, since he has become a mortal.

The affirmation of the nothingness of being is not the only sign that metaphysics grows on the solitude of the earth. Metaphysics is the explicit affirmation that the earth is the content of immediate knowing— and this is an explicit affirmation of the earth's solitude. The earth is in fact *ta phusika*, for *phusis* is the region of Becoming and reality-that-becomes is the totality of what immediately appears. Precisely because it "draws from Becoming to Being" (*holkon apo tou gignomenou epi to on*, Plato, *The Republic*, 521d), metaphysics posits the region of Becoming as what, certainly, must be transcended, but which, for this very reason, is the indubitable dimension from which knowing must proceed (and with which man, in his present life, originally deals). This is the fundamental property of every type of metaphysics, whether *phusis* be understood as a being outside the mind or as the content of the *cogito* or of phenomenological description; whether metaphysics, in transcending the region of Becoming, comes to affirm an immutable being (a being which transcends Becoming) or comes to identify reality-that-becomes with the totality of Being, positing the original content of experience in the form of thought. The Indo-European root of *kosmos* is *kens*; "to announce with authority" (Latin *censeo*). For metaphysical thought *kosmos* is *phusis* understood as the region of Becoming; which means that the earth is the sure place, the region that announces itself with authority (and silences the voices of myth). The "world" is the earth as the sure place and so as solitude. In the untruth of premetaphysical man, the nothingness of the earth is the invisible thought that (countered by truth) guides his every step. This thought leaves its traces in man's works, but metaphysics alone has testified to it and made it visible. Metaphysics is thus the uncovered trace of the fall of man; and the history of the West is that dizziness from the fall which is the West's awareness of its own dominant thought.

The West, in receiving the earth, enveloped it in solitude and fell into untruth. Metaphysical dominance is the trace of the fall. This trace is lacking in the works of nonmetaphysical peoples, where all testimony to the truth of Being is silent. Their works too (like all works) preserve traces of truth—which stands eternally gathered before *all* peoples—but they do not testify to it. We do not know how other peoples received the earth. We may suppose that they, too, whose works testify only to earthly things, saw and experienced the earth as the sure ground—as the only thing that *can* be testified to—and therefore that they, too, fell into the untruth of Being. It may be supposed that the fall is part of the essence of all peoples. But these suppositions still cannot be evaluated. Do only mortal peoples inhabit the earth?

12. The Earth's Isolation and the "Parricide"

Western civilization is the only testimony in man's history to the truth of Being. But the West addressed itself to the truth of Being while grounding itself on the solitude of the earth, and truth's only testimony became its most abysmal betrayal. Isolation, which nihilates the earth, determines the relation that metaphysics establishes between the things of the earth and their *Being*. Throughout the course of its history, metaphysics has attempted to think the Being (the existence) of what is originally seen as a Nothing. Seeing the nothingness of things in fact means positing them as isolated from their Being and recognizing the essential accidentality of their relation with Being (i.e., of their existing). It is therefore inevitable that, while explicitly opposing being to Nothing, metaphysics also comes to explicitly affirm that being is nothing (when it is not and insofar as it can not-be).

The earth's solitude envelops also he who first named the truth of Being. All antiquity attests that Parmenides affirmed pure Being, while denying the existence of the determinations of the manifold. Acting at the root of this negation is the absolute separation—the isolation—of determinations from Being.[4] Isolated from Being—thought, that is, in their separation from Being and so from the truth of Being—determinations must necessarily be understood as a Nothing. Parmenides posits them as a Nothing, not because he does not yet know the Platonic distinction between *heteron* and *enantion*, but precisely because he isolated them from Being; and thus, in isolation, the *heteron* must be posited as the *enantion tou ontos*. But the determinations of the manifold are in the first place the determinations-that-become of the earth, and their separation from pure Being—that is, from that of the truth of Being to which Parmenides did testify—expresses the way in which Parmenides keeps the earth isolated from the truth of Being. Thus the way in which the manifold things of the earth are thought (the way of solitude) determines the way in which the manifold in general is thought. Hence it is *precisely because* Parmenides *too* is convinced that the earth is the sure ground that, when he measures the earth against the trait of the truth of Being to which he had testified— and this trait is the pure eon whose dazzle tries to ravish the witness from the earth—he is compelled to posit the untruth (*ouk eni pistis alēthēs*, frag. 1, v. 30), the illusoriness, the unsureness, and, ultimately, the nothingness of the earth.

And the very link with which Plato unites Being to its determinations is forged in the solitude of the earth. In its truth, Being is not pure Being, but rather the union of pure Being and a determination. Plato is the witness to this union, but he unites to Being what also for him is originally understood as absolutely isolated from it. Plato, too, isolates the earth (and then the totality of determinations) from Being, and therefore also Plato must take the earth and, in general, determinations, to be a Nothing. It is precisely this Nothing—that is, this non-Nothing which, isolated from Being, must be posited as a Nothing—it is precisely this non-Nothing, now understood as nothing, that he unites to Being (i.e., to its being a non-Nothing). For Parmenides Being is the pure "is"; a determination (such as "house"), isolated from its "is," must be posited as a Nothing. Plato, in contrast, knows that Being is a determination-that-is (e.g., a house-that-is), since "is" means "is not a Nothing," and a determination, for example, "house," is not a Nothing (is not a meaning-nothing). But in forging this link between a determination and its "is," Plato—like Parmenides before him—from the outset isolates the determination from its Being and so, from the outset, has to understand it as a Nothing. Thus he unites to Being (i.e., to not-being-a-Nothing) what is destined to be thought as a Nothing. Insofar as a determination is already thought as a Nothing, metaphysics deems legitimate the accidentality of its union with Being. Once this has been admitted, it cannot but be affirmed that a determination "is when it is, and when it is not, it is not," and that therefore its coming-to-be is a process in which it (the non-Nothing!) has been, and returns to being, a Nothing. Even the Idea (*ousia ontōs ousa*—and each of the gods of the West), *qua determination*, is a Nothing. In fact, in order to posit it as Being—as that Being whose fate is never to be a Nothing and which therefore is *ontōs on*—Plato and all Western thought must resort to *reasons* that are different from the *true*—and unthought—reason, which is the *truth* of the earth. If the earth comes out of solitude and determinations are no longer isolated from their Being and from the truth of their Being, then it is determinations *as such*—*any* determination of Being—that must be posited as *ontōs on*, that is, as that which can never have been, nor ever return to being, nothing. God—in contrast—is the result of the will to posit as Being (*ontōs on*) what is originally thought as a Nothing.

The dominant thought of metaphysics is the identity of Being and Nothing—yet metaphysics explicitly undertakes to safeguard and preserve their opposition. The Hegelian dialectic is one of the paramount forms of metaphysical thought; but the identity of Being and Nothing, which constitutes the first triad of Hegelian logic, is by no means a formulation of the dominant thought of metaphysics. Hegel in fact stresses that the identity of Being and Nothing is not the identity of determinate Being (*Daseyn*) and Nothing (as if it were "the same whether I am or am not, whether this house is or is not, whether these hundred talers are, or are not, part of my fortune"[5]), since that which is identical to nothing is *pure* Being, Parmenides' pure "is," isolated from determinations. Therefore, according to Hegel, common sense has no call to be astonished at the identity of Being and Nothing (it would, rather, have good reason to be astonished at their difference, as in fact Friedrich Adolf Trendelenburg was); and therefore it is not a matter of indifference, according to Hegel, whether something determinate is or is not.[6] Like Plato and Aristotle, Hegel defends the noncontradictoriness of being: he too undertakes to safeguard the opposition of not-Nothing and Nothing. But for this very reason, Hegel, too, identifies Being and Nothing: not in the sense of the first triad of the *Logic*, but rather in the same sense in which Plato and Aristotle do so. In the very act in which it affirms the opposition of being and Nothing, metaphysical thought allows being to be a Nothing. Hegel opposes being (*Daseyn*) to Nothing, but he distinguishes finite beings, whose destiny is to be born and perish—to have been, and return to being, nothing—from privileged being, which is itself the eternal becoming of the finite (so that the nothingness of finite being is the condition of the eternity of privileged being). Here too, the determinate, as such, issues from and returns to nothingness, because the basis of Hegelian metaphysics, too, is the isolation of the earth—and, therewith, of determination as such—from the truth of Being.

The fundamental metaphysical doctrine, designed to clarify the meaning of the isolation of determinations, is the Hegelian doctrine of abstract understanding. And yet, this epic struggle against the isolation

of determinations is guided by a thought that is completely enveloped by
the solitude of the earth. Hegel's *Logic* intends to be the overcoming of
the abstractness of pure Being (the "is"), which, as isolated from determinations, is (in its turn) a Nothing. Indeed, dialectical development is the
determinate mode according to which the synthesis between pure Being
and the totality of determinations is instituted. But the ultimate meaning
of this Hegelian synthesis is still the Platonic one. From the outset a bot-
tomless abyss (the abyss which isolates the earth from the truth of Being)
yawns between pure Being and determinations, so that the union of the
two sides is the synthesis of what from the outset was destined to remain
divided. In dialectical development, pure Being determines itself (i.e., it
unites with determinations). But the determination—as *empirical*, and not
privileged, determination—maintains in its synthesis with Being the
character that from the outset belongs to it as separated from Being—
namely, the character of being a Nothing. It is united to Being but continues to be a Nothing. Thus it is inevitable that Hegel should treat the
determination as a Nothing and affirm its synthesis with Being to be accidental, and that therefore it is destined to be born and to perish (i.e.,
to have been, and return to being, nothing)—only that privileged being,
which is the very accidentality of the synthesis, remaining eternal. (In
Hegel, the synthesis between pure Being and *categorical* determinations is
indeed intended to count as necessary and not as accidental. But this
necessary synthesis—the organism of categories, the Idea—is once again
the privileged structure that, as in Plato, makes the becoming of empirical
determinations—i.e., the institution of the accidental synthesis between
the categorical and the empirical—possible.)

Aristotle had already reproached Parmenides with isolating Being
from determinations (*Physics* 186a22), stressing that Being is other than
determinations as *distinct* (*tōi emai heteron*), and not as *separate* from them
(*ou gar ēi khōriston*). Distinctness does not imply the nothingness of determinations ("*outhen hētton polla ta leuka* [= *polla ta onta*] *kai oukh hen*"), precisely because what is other than Being is distinct from it in meaning but
is not something isolated from it ("*allo gar estai to einai leukōi* [= *onti*] *kai to
dedegmenōi, kai ouk estai para to leukon* [= *on*] *outhen khōriston*"). And yet, the
isolation of determinations with which Aristotle reproaches Parmenides
underlies Aristotelian—and Platonic, and Hegelian—metaphysics. Precisely because determinations isolated from Being (and from the truth of
Being) are a Nothing, metaphysical thought can admit that, coming to
be, they have been nothing (and return to being nothing): "For what is
generated is what is-not" (*gignetai gar to mē on*, Metaph. 1067b31). Indeed,
for metaphysics the greatest difficulty lies in thinking that a thing is not
a Nothing.

13. *The Salvation of Truth*

Pure contradiction is the original dwelling place of man. The eternal manifestation of the truth of Being is the primordial structure of

contradiction. Any other contradiction is grounded on this structure. Coming-out of primordial contradiction is an occurrence, since such contradiction is the attitude that has always been assumed by the never-setting background, which is the place where every occurrence can be received. The "life of the gods," which the peoples of the earth lead in their original dwelling places, therefore has always awaited its salvation. Salvation lies in the occurrence. The true meaning of salvation—that is, the truth of salvation—is in fact the salvation of truth. Since man is the eternal guardian of the truth of Being, the true meaning of the salvation of man is the salvation of truth. And for truth salvation means passing beyond the contradiction that has always penetrated it.

The occurrence is a coming-out of the motionless unrest of the never-setting. One does not come out of, that is, one does not escape, the never-setting; instead, one comes out of the primordial contradiction in which the never-setting finds itself, owing to its not being the completed manifestation of every trait of Being. Although the never-setting does let every being appear in the "whole," it cannot bring out into the light all that concrete richness of things which the meaning of "whole," by appearing, demands be brought out (and thus what is not the whole is made to mean "whole"). Since the never-setting background is that without whose appearing nothing could appear, contradiction does not belong to the background in the sense that, without contradiction, nothing could appear. For contradiction is the attitude assumed by the background insofar as it does not contain the concrete whole of Being. Contradiction is not that without which nothing could appear, but rather the background is invested with contradiction owing to the not-appearing in it of the whole. Salvation is the completion of the revelation of Being which is granted to the eternal appearing of the truth of Being.

The occurrence is unique. If in addition to the earth something else occurred—heaven, the beyond—then both would constitute the content of the occurrence. Therefore the occurrence occurs in the occurrence of the earth; and salvation lies in the occurrence. But the occurrence is also the greatest of perils—the possibility of abysmal perdition. Being offers the earth to the guardians of the truth of Being. And the offering was accepted. The receiving of the offering belongs to the offering: it is the way in which the occurrence occurred. Receiving the offering means willing the continuation of the occurrence to the completion of its occurring. Its total completion gives the measure of the disclosure of Being which is granted to the essence of man, thus giving the measure of the liberation from contradiction: the measure of salvation. But is the earth the measure granted, or is this measure given by a different completion of the occurrence?

The guardians of truth received the offering while leaving it in solitude. We have no means of shedding light upon man's fall into the solitude of the earth. But insofar as the earth's solitude is a fact, it has no need of light. Since the truth of Being eternally appears, man's life in untruth is possible only as a conflict, in Appearing, between the truth of Being and the isolation of the earth. The distraction from truth is manifest not only in our everyday existence (which seems concerned with anything but the truth of Being) but also in Western civilization itself—the civilization that today rules the earth. Yet it could not appear if isolation had not occurred. But why did the earth's peoples, when they received the offering, envelop it in solitude? Any answer to this question is still only a posible interpretation.

The alienation opposed to that in which the peoples of the earth have fallen is the rejection of the earth, that is, the will that the occurrence should not continue. No trace has been uncovered of this form of alienation. Suicide (which in our culture has become a form of metaphysical nihilism) is one of the events that occur within the receiving of the earth. Rejection of the earth is a form of alienation, because only the occurrence can bring salvation. If the isolation of the earth is a negation of the truth of Being, so is the rejection of the earth: for the rejection of the earth is the rejection of salvation. Salvation may be rejected, because in the occurrence which brings it one fears abysmal alienation. One may enclose the earth in solitude, because it is believed that the only salvation possible lies in the way in which the occurrence of the earth stands before us. If salvation disappoints, one attempts the supreme feat of forgetting the *ground* of the disappointment—namely, the truth of Being. And the only way in which man can do so is by isolating the earth. But how can the value of this interpretation be established?

In fact, truth—as testified to in philosophy—does not even know whether the history of salvation is a necessary development to the never-setting background of Appearing (i.e., an epiphany of Being, which, like the appearing of the background, is ineluctable), or whether it is a history of freedom. The offering of the earth, the receiving of the offering, the fall into solitude, the metaphysical alienation of the West, and all the ways of the occurrence: are these the steps of freedom or of inevitable necessity? And the truth of Being, which eternally appears—what does it know of the measure of its own salvation? For even if, in philosophy, the truth of Being does remain here before us with the setting of the isolation of the earth, we do not know whether, in testifying to it, we are testifying to the whole that eternally appears.

14. *Repetition of the Acceptance of the Offering*

And yet philosophy, as the guardian of the truth of Being, is the repetition of the supreme moment of the history of salvation. The slow reflux which in philosophy carries the earth's solitude to its setting places the truth of Being anew before the offering of the earth and allows it to repeat the receiving of the offering. In the primordial receiving, the earth's peoples brought the earth to encroachment. Their fall into untruth made the earth itself a work of untruth. The West has become the leader of untruth, its bearer and dominant witness; and so the earth has become a work of the West. Yet the only testimony to the truth of Being is in the history of the West. The possibility of repeating the receiving of the offering was granted to Greek thought. And in fact the Greeks came close to the repetition. Greek thought looked out on the testimony to the truth of Being, but without stopping there it continued its course, leading the West along the path of Night, into the remotest distance. Will the West's wandering star approach that testimony anew, making it possible to repeat the receiving of the earth?

In philosophy, the truth of Being again encounters the offering of the earth. Untruth's rampant dominance is crossed by a reflux which slowly carries it toward its setting and allows the truth of Being to reemerge. The earth places itself anew, uncontended, before the eyes of truth, whose guardian is philosophy. In fact, the truth of Being is testified to in philosophy, because the conviction which isolated the earth and contends for it against truth has set. But the earth remains before us laden with the fruits of its long solitude. Truth calls it back from exile, but the voice of truth now finds it in various guises, for the earth is laden with all the time and all the works of alienation. And yet in philosophy the possibility that the history of the salvation of peoples should begin anew is safeguarded. We are faced with the supreme test, on which the completion of the occurrence and the conclusion of the history of salvation depends. Thought that testifies to the truth of Being may again be swallowed up by the solitude of the earth, and the peoples of the earth may definitively move away from the truth of Being. But in the possibility that opens all hope lies.

Untruth is such, not because it wills the earth and the earth's continuation, but because it isolates what is willed. This willing, as such, is the same original will with which the eternal appearing of the truth of Being wills the earth and its continuation. The earth's peoples accepted the offering. This is affirmed, precisely because the earth and its continuation in fact appear as willed. The repetition of the receiving of the offering is therefore not a new occurring of what has occurred ever since Being offered the earth to man. Ever since the offering occurred, the earth has appeared in the truth of Being, which since then has also been the truth of the earth. Untruth is not a conflict between the pure truth of Being (to which the earth is not yet linked) and the isolation of the earth, but rather between the truth of Being, which is also the truth of the earth, and the isolation of the earth. Thus it is a conflict between the truth of the earth, which wills the continuation of the earth in truth, and the isolation of the willed earth, in which willing becomes the will that the earth should continue in solitude. The repetition of the receiving of the offering is, therefore, the setting of the conviction that isolates the earth; so that, with this setting, not only do we remain what we have always been, but we also remain what we have begun to be ever since the offering of the earth occurred—we also remain the receiving of the offering.

Philosophy is not the return of the silence of man's original dwelling place, because philosophy preserves the occurrence of the earth in the truth of Being; but with the setting of the isolating conviction, the earth is called by the pure voice of truth. Yet the earth—unsetting work of the West's untruth—is indifferent to the voice of truth. The voice of solitude sets, but the works it called forth do not, and ever more vertiginous is the West's race towards the constellations of Night. The earth, as a work of the West's alienation, appears in the truth of Being; so that philosophy— insofar as it is our remaining what we have always been and what we have begun to be since the offering of the earth occurred—is the contradiction in which solitude and nihilism are superseded (i.e., in which they set) only in the abstract element of thought, while their works are left as not-superseded. This contradiction is the way in which man is faced with the supreme test, to which the completion of the history of salvation is linked. Since the true meaning of salvation is the salvation of truth—that is, the completion of that revelation of Being which is granted to the eternal appearing of the truth of Being—any other meanings of salvation can be accepted only if they can be conjoined with this original meaning. Even if we allow that the kerygma can save in this original sense, theology, especially today, realizes that the conditions of hearing the kerygma are lacking. But also theology is dominated by metaphysical nihilism: its accusation that this hearing is impossible in our time is grounded on that same dominant thought—that is, the nothingness of being—which itself prevents true hearing. The only hearing in the history of the West has been metaphysical hearing, in which everything is made to pass through the solitude of the earth and where, therefore, no Advent and no kerygma can bring salvation.

Philosophy, insofar as it witnesses the setting of the isolating conviction, makes true hearing possible. But if salvation is to occur, the road to salvation must pass through the setting of the works of solitude and so through the setting of the West, the dominant witness to solitude. Awaiting this setting, philosophy looks out on the supreme possibility of the peoples of the earth. Only if, in philosophy, the isolating conviction does not counter the truth of the earth, can the earth be brought to setting as a work of untruth. For the earth to become a work of truth, it is first necessary that the truth of the earth should not appear countered by the isolating conviction. It is for this reason that philosophy brings us back to the fork in the road which once opened up before the original dwelling place of man: to the right, the untrodden path of Day, where the earth becomes a work of the truth of Being; and to the left, the path of Night, which leads the earth into solitude. Is philosophy the dawn of Day? Or is it the swan song before the truth of the earth is definitively caught up in its conflict with the isolating conviction and the West resumes its precipitous course, never looking back?

Nor can the individual save himself independently of the history of the West, that is, of the way in which the earth's peoples bring to completion the occurring of the earth. As long as philosophy is the contradiction in which the isolating conviction sets but the works of solitude do not, no *eupraxia* can bring salvation: it falls on sick ground and becomes sick itself. Any single individual's resolve to save himself independently of the configuration that historical objectivity assumes, will be merely pathetic. But insofar as the individual is the guardian of the truth of Being, he is the good shepherd who calls his peoples back to the parting of the ways and shows them the path of Day.

Taking this path—bringing the earth as a work of the West to its setting—means superseding the contradiction of philosophy and so bringing philosophy toward its completion. Philosophy is contradiction insofar as it is only our remaining what we eternally are and what we have begun to be since the earth was offered to us. But on the path of Day philosophy is the earth, which becomes the uncontested work of truth. And so it is the completion of the occurrence: Being's assent to the will that the occurrence continue and be accomplished in the truth of Being.

TRANSLATED BY GIACOMO DONIS

Contributors

GIOVANNA BORRADORI teaches aesthetics at the Faculty of Architecture, Milan Polytechnic. Her publications include *Il Pensiero Post-Filosofico* (1988) and articles in such journals as *Rivista di Estetica, Alfabeta, The Journal of Aesthetics and Art Criticism,* and *Social Text.* She is currently translating Jacques Derrida's *Memoires pour Paul de Man* into Italian.

MASSIMO CACCIARI first proposed a conjugation between the dialectical tradition of Marxism and the nihilistic horizon in which the ontological discourse of Nietzsche and Heidegger operate. His publications include *L'angelo necessario* (1985).

UMBERTO ECO is a semiotician and best-selling fiction writer. Among his books published in English are *The Role of the Reader: Explorations in the Semiotics of Texts* (1979) and *Semiotics and the Philosophy of Language* (1984).

ALDO G. GARGANI, a distinguished Wittgenstein scholar, edited and translated Wittgenstein's *Memoires* into Italian. His recent writings have appeared in journals and anthologies such as *Filosofia '86.*

MARIO PERNIOLA is a leading figure in the Italian contemporary debate who first raised the issue of "the statute of the referent," which led to the discussion of "dissimulation" and "simulacra." His books include *Transiti: come si va dallo stesso all stesso* (1985).

FRANCO RELLA is a scholar of German and Austrian avant-garde art of the turn of the century. He has focused his work on the connection between the aesthetics of modernism and the culture of the Vienna Succession, including Musil, Rilke, Wittgenstein, and Kafka. His publications include *La battaglia della verità* (1986) and *Metamorfosi. Immagini del pensiero* (1984).

PIER ALDO ROVATTI, a crucial figure within Italian post-Marxist debate, has more recently devoted his interests to the relationship between the "weak" hermeneutical perspective and French post-structuralist theories, with particular attention to Derrida, Lévinas, and Lacan. He is the co-editor, with Gianni Vattimo, of *Il Pensiero Debole* (1983).

EMANUELE SEVERINO has animated recent Italian philosophical discussion with his attempt at accomplishing a radicalization/overcoming of Heidegger's critique of Western metaphysics. His publications include *La struttura necessaria* (1979) and *Essenza del Nichilismo* (1982).

GIANNI VATTIMO has recently become a leading figure in the contemporary philosophical debate due to his theory of "weak thought." His publications include Il *soggetto e la maschera* (1981), La *fine della modernità* 1985, and the anthology Il *pensiero debole* (1983), co-edited with Pier Aldo Rovatti.

NOTES

Introduction: Recoding Metaphysics

1) Many anthologies dedicated to continental philosophy do not even mention the Italian contribution. Particularly significant cases are those devoting all their attention to topics that represent the central nodes of Italian debate since World War II, such as phenomenology, existentialism, and hermeneutics. See, e.g., *Continental Philosophy in America*, ed. Hugh Silverman, John Sallis, and Thomas Seebohm (Pittsburgh: Duquesne Univ. Press, 1983), focused on Husserl, Heidegger, and Merleau-Ponty and the impact of their thought on contemporary research. Also *Hermeneutics and Deconstruction*, ed. Hugh Silverman and Don Ihde (Albany: State of New York Press, 1985); and *Hermeneutics and Praxis*, ed. Robert Hollinger (Notre Dame, Ind.: Univ. of Notre Dame Press, 1985), both concerned with the role of hermeneutics in contemporary philosophy and criticism. An important problem concerning the lack of knowledge about the Italian scene has to do with its very "philosophical" specificity, especially in comparison with the latest French positions of structuralism and poststructuralism. In fact, many French authors (from Barthes to Foucault, Deleuze, and Derrida) have been contextualized in America within the broad field of "literary studies." For this reason, exhaustive presentations of the structuralist/poststructuralist debate do not take any Italian contribution into consideration. See, e.g., *Textual Strategies: Perspectives in Post-Structuralist Criticism*, ed. Hosué V. Harari (Ithaca, N.Y.: Cornell Univ. Press, 1979).

2) For the history of Italian philosophy, see Eugenio Garin, *Storia della filosofia italiana*, 3 vols. (Turin: Einaudi, 1966).

3) In order to contextualize the contemporary debate within the broader perspective of the history of Italian philosophy since World War II, see the volumes of collected essays *La filosofia italiana dal dopoguerra ad oggi* (Bari: Laterza, 1985), and *La cultura filosofica italiana dal 1945 al 1980* (Naples: Guida, 1982).

4) For a survey of this hermeneutical position of Italian contemporary philosophy see Giovanna Borradori, "Weak Thought and Postmodernism: The Italian Departure from Deconstruction," *Social Text* 18 (Winter 1987/88): 39–49. A very good critical reading of this tendency, within its proper historical perspective, is given in the essay by Valerio Verra, "Esistenzialismo, fenomenologia, ermeneutica, nichilismo," in *La filosofia italiana dal dopoguerra ad oggi*. A manifesto of contemporary Italian hermeneutics is the anthology *Il pensiero debole*, ed. Gianni Vattimo and Pier Aldo Rovatti (Milan: Feltrinelli, 1983; Baltimore: John Hopkins Univ. Press, forthcoming). Other volumes of collected essays that are emblematic of this debate are *La crisi della ragione*, ed. Aldo G. Gargani (Turin: Einaudi, 1979); and *Filosofia '86*, ed. Gianni Vattimo (Bari: Laterza, 1987).

5) A useful contextualization of this "radical" side of the Italian debate within its proper historical perspective is given by Adriano Bausola in his essay "Neoscolastica e spiritualismo," in *La filosofia italiana dal dopoguerra ad oggi*. Only in some respects did Massimo Cacciari contribute to developing this theoretical position, particularly with his book *Krisis: Saggio sulla crisi del pensiero negativo da Nietzsche a Wittgenstein* (Milan: Feltrinelli, 1976). Other basic references to this side of the Italian debate are the following works by Emanuele Severino: *La struttura originaria* (1958; Milan: Adelphi, 1981), *Il destino della necessità* (Milan: Adelphi, 1980), and *Essenza del Nichilismo* (Milan: Adelphi, 1982).

6) Literary journals have attempted to present the hermeneutical side of the Italian contemporary debate by dedicating entire issues to this topic. See *Differentia* 1 (Autumn 1986), and *Substance* (Fall 1987). In French see *Les philosophes italiens par eux-mêmes*, Critique, nos. 452–53 (Jan.–Feb. 1985).

7) This difficulty is due to two factors: In the first place, a basic lack of knowledge about the Italian scene itself, which makes it impossible to get a historical perspective in which to contextualize contemporary philosophical events. In the second place, the failure of Italian culture to exorcise the experience of Fascism, which, in order to pursue its nationalist politics, put a strong emphasis in every field on the "national" element. This is why, still today, many authors hesitate acknowledging the national roots of their discourses, in which, very often, their originality lies. An emblematic case of this "anti-Fascist censorship" is the discussion (which will be more deeply explored later on) of the role Vico and Croce played, and still play, in the development of Italian philosophy. An accurate historical account of the influences of historicism throughout twentieth-century Italian thought is the essay by Eugenio Garin, "Agonia e morte dell'idealismo italiano," in *La filosofia italiana dal dopoguerra ad oggi*.

8) In his exhaustive essay "Il carattere della filosofia italiana contemporanea," in *La cultura filosofica italiana dal 1945 al 1980*, Carlo Augusto Viano points out the important role played by the "eclectic strategy" in the constitution of Italian cultural unity during the Risorgimento. "Since its origins, that is, after Napoleonic resettlement of Europe, Italian philosophical culture, very sensitive to its own peculiarity, had to face the perspective of borrowing from foreign cultures and of reconsidering its own tradition, more or less remote in time. On the one hand, the image or mirage of an Italian tradition which through Vico reached back to the Renaissance held sway. On the other hand, one glimpsed the presence of European culture which could not easily be taken back to that tradition" (23).

9) Vattimo and Rovatti, "Premise" to *Il pensiero debole*, 10.

10) Giambattista Vico, *The New Science of Giambattista Vico*, trans. Thomas G. Bergin and Max H. Fisch (New York: Cornell Univ. Press, 1970).

11) Vattimo, "Dialettica, differenza, pensiero debole," in *Il pensiero debole*, 27.

12) It is crucial to remember that by "economics" Croce meant a discipline-container in which flow into each other all of the disciplines oriented toward practical success or technological efficiency, all experimental sciences, or more generally, all sciences proceeding by generalization.

13) That these "internationalist openings" refused Croce's legacy and even fought against it is in itself contradictory. As a matter of fact, their strategy was to operate on different philosophical systems trying to adapt them to the Italian situation, forgetting about their original context and manipulating them freely as if they were abstract historical objects, a typical historicist attitude. Even though throughout twentieth-century philosophy, there had been cases of commingling between different systems (such as between Marxism and psychoanalysis in Germany and France between the two world wars, and between Marxism and existentialism in France after the Second World War), what distinguished the Italian situation of the 1950s was the degree to which these comminglings took place. Marxism was wed not only to existentialism, but also to pragmatism, neopositivism, and phenomenology. Moreover, such an intersection of different philosophical systems started out from an abstract criticism of each individual formulation: to this extent, neopositivism was considered a global theory of rationality and not, as originally, a criticism of scientific language. In the same way, the metaphysical élan and connection to specific aspects of American society of American pragmatism was ignored, so that it appeared as a "behavioristic" pedagogical theory. As Viano points out ("Il carattere della filosofia italiana contemporanea," 21–23), the "eclectic strategy" of Italian postwar philosophy would never have existed without the historicist background.

14) Enzo Paci, *La filosofia contemporanea* (Milan: Garzanti, 1957), 65.

15) See Enzo Paci, *Ingens sylva: Saggio su Vico* (Milan: Mondadori, 1949); and Paci, *Esistenzialismo e storicismo* (Milan: Mondadori, 1950).

16) Enzo Paci, *Il nulla e il problema dell'uomo* (Turin: Taylor, 1950); and Paci, *Dall'esistenzialismo al relazionismo* (Florence: D'Anna, 1957).

17) Enzo Paci, *Funzione delle scienze e significato dell'uomo* (Milan: Mondadori, 1963), trans. as *The Function of Sciences and the Meaning of Man* by Paul Piccone

and James Hansen (Evanston, Ill.: Northwestern Univ. Press, 1972); and Paci, *Idee per un'enciclopedia fenomenologica* (Milan: Bompiani, 1973).

18) See Vattimo, "Dialettica, differenza, pensiero doble"; Aldo G. Gargani, "L'attrito del pensiero"; and Pier Aldo Rovatti, "Tenere la distanza," the latter two translated into English in this volume as "Friction of Thought" and "Maintaining the Distance," respectively.

19) See Vattimo and Rovatti, "Premise" to *Il pensiero debole*, 9.

20) Luigi Pareyson, "Federico Guglielmo Giuseppe Schelling" and "Giovanni Amedeo Fichte," in *Grande Antologica Filosofica* (Milan: Mondadori, 1954).

21) Luigi Pareyson, *La filosofia dell'esistenza e C. Jaspers* (Naples: Loffredo, 1950); Pareyson, *Esistenza e persona* (Genoa: Il Melangolo, 1985); and Pareyson, *Estetica: Teoria della Formatività* (Florence: Sansoni, 1974).

22) See Armando Rigobello, *L'impegno ontologico: Prospettive attuali in Francia e riflessi nella filosofia italiana* (Rome: Armando, 1977).

23) Luigi Pareyson, "Rettifiche sull'esistenzialismo," in *Studi di filosofia in onore di G. Bontadini* (Milan, 1975), 246–47.

24) Gianni Vattimo, "Verso un'ontologia del declino," in *Al di là del soggetto* (Milan: Feltrinelli, 1981), 51; translated in this volume as "Toward an Ontology of Decline," 63.

25) Vattimo, *Al di là del soggetto*, 13.

26) Ibid.

27) See Gianni Vattimo, *Il soggetto e la maschera: Nietzsche e il problema della liberazione* (Milan: Feltrinelli, 1974). An interesting overall glance at the Italian debate on Nietzsche can be found in the appendix to Friedrich Nietzsche, *Il libro del filosofo*, ed. M. Beer and M. Ciampa (Rome: Armando, 1978).

28) See Cacciari, *Krisis*; also, Massimo Cacciari, *Pensiero negativo e razionalizazione* (Venice: Marsilio, 1977).

29) See *La crisi della ragione*, ed. Gargani, 30.

30) Umberto Eco, "An *Ars oblivionalis*? Forget It!" trans. Marilyn Migiel, PMLA 103, no. 3 (May 1988): 255.

31) Ibid., 259.

32) Umberto Eco, "Intentio lectoris: The State of the Art," 39–40, in this volume.

33) Gustavo Bontadini, Dal problematicismo alla metafisica (Milan: Marzorati, 1952).

34) Emanuele Severino, "La terra e l'essenza dell'uomo," in Essenza del Nichilismo, 216; translated in this volume as "The Earth and the Essence of Man," 177.

35) Severino, "Premise" to La struttura originaria, 16.

36) Ibid., 89.

37) See Emanuele Severino, "Che cosa significa pensare" and "Le necessità dell'Occidente e la Necessità", pars. 10–11 of the "Premise" to La Struttura originaria, 90–98.

Intentio Lectoris: The State of the Art

Booth, Wayne. The Rhetoric of Fiction. Chicago: University of Chicago Press, 1961.

Chatman, Seymour B. Story and Discourse: Narrative Structure in Fiction and Film. Ithaca, N.Y.: Cornell Univ. Press, 1978.

Corti, Maria. Principi della comunicazione letteraria: Introduzione alla semiotica della letteratura. Milan: Bompiani, 1976.

Derrida, Jacques. Of Grammatology. Translated by Gayatri Chakravorty Spivak. Baltimore: Johns Hopkins Univ. Press, 1976.

Eco, Umberto. The Role of the Reader: Explorations in the Semiotics of Texts. Bloomington: Indiana Univ. Press, 1979.

————. Semiotics and the Philosophy of Language. Bloomington: Indiana Univ. Press, 1984.

————. Theory of Semiotics. Bloomington: Indiana Univ. Press, 1976.

Eco, Umberto, and Thomas Sebeok. The Sign of Three. Bloomington: Indiana Univ. Press, 1983.

Foucault, Michel. "What is the Author?" in Language Counter-Memory Practice, edited and translated by Donald F. Bouchard, 113–38. Ithaca, N.Y.: Cornell Univ. Press, 1977.

Genette, Gerard. *Figures* III. Paris: Seuil, 1972.

Hirsch, E. D. *Validity and Interpretation.* New Haven, Conn.: Yale Univ. Press, 1967.

Holub, Robert C. *Reception Theory: A Critical Introduction.* New York: Methuen, 1984.

Iser, Wolfgang. *The Implied Reader: Patterns of Communication in Prose Fiction from Bunyan to Beckett.* Baltimore: Johns Hopkins Univ. Press, 1974.

————. *The Act of Reading: A Theory of Aesthetic Response.* Baltimore: Johns Hopkins Univ. Press, 1978.

Jauss, Hans Robert, ed. *Nachahmung und Illusion.* Kolloquium Giessen, June 1963. Munich: W. Fink, 1969.

Kristeva, Julia. *Le texte du roman: Approche sémiologique d'une structure discursive.* The Hague: Mouton, 1970.

Lotman, Juri. *The Structure of the Artistic Text.* Translated by Ronald Vroon. Ann Arbor: Univ. of Michigan Press, 1977.

Miller, J. Hillis. *Thomas Hardy: Distance and Desire.* Cambridge, Mass.: Harvard Univ. Press, 1970.

————. "On Edge: The Crossways of Contemporary Criticism." *Bulletin of the American Academy of Arts and Sciences* 32, no. 4 (1979): 13–32.

Pratt, Marie Louise. *Toward a Speech Act Theory of Literary Discourse.* Bloomington: Indiana Univ. Press, 1977.

Riffaterre, Michele. *Essai de stylistique structurale.* Introduced by Daniel Delas. Paris: Flammarion, 1971.

Rorty, Richard. *Consequences of Pragmatism: Essays, 1972–1980.* Minneapolis: Univ. of Minnesota Press, 1982.

1. I realize now that my idea of system of expectations, though built up on the grounds of other theoretical influences, was not so dissimilar from Jauss's notion of *Erwartungshorizont.*

2. In *Opera aperta* I was considering under the heading "work of art" not only literary texts but also paintings, cinema, television. I am grateful to Wolfgang Iser (1978) for observing not only that some of my remarks on nonverbal arts were also relevant for literature (chap. 5), but also (chap.

3) that my further discussion on iconic signs (Eco 1968) supported the idea that even literary signs designate "the conditions of *conception* and *perception* which enable the observer to construct the object intended by the sign" and therefore "constitute an organization of signifiers which do not serve to designate a signified object, but instead designate the *instructions* for the *production* of the signified."

3. One could say that, while the semantic reader is planned or instructed by the verbal strategy, the critical one is such on the grounds of a mere interpretative decision—nothing in the text appearing as an explicit appeal to a second-level reading. But it must be noticed that many artistic devices, for instance, stylistic violation of the norm or defamiliarization, seem to work exactly as self-focusing appeals: the text is made in such a way as to attract the attention of a critical reader. Moreover, there are texts that explicitly require a second-level reading. Take for instance Agatha Christie's *The Murder of Roger Ackroyd*, which is narrated by a character who, at the end, will be discovered by Poirot as the murderer. After his confession, the narrator informs the readers that, if they had paid due attention, they could have understood in which precise moment he committed his crime because in some reticent way he did say it. See also my analysis of Allais's "Un drame bien parisien" (Eco 1979), where it is shown how much the text, while step-by-step deceiving naive readers, at the same time provides them with a lot of clues that could have prevented them from falling into the textual trap. Obviously, these clues can be detected only in the course of a second reading.

Metaphysics, Violence, Secularization

This discussion of the problem of going beyond metaphysics is related to several of my recent essays: "La secolarizzazione della filosofia," *Il Mulino*, no. 300 (1985); "Ermeneutica e secolarizzazione: A proposito di L. Pareyson," *Aut Aut*, no. 213 (1986); "Ritorno alla (questione della) metafisica," *Theoria* (1986).

1. Friedrich Nietzsche, *Human All-Too-Human*, Part II, trans. Paul V. Cohn, ed. Oscar Levy (1909–11; rpt. New York: Russell & Russell, 1964), 239.

2. Pierre Klossowski, *Nietzsche et le cercle vicieux* (Paris: Mercure de France, 1969).

3. See Martin Heidegger, *Being and Time* (New York: Harper & Row, 1962), par. 32.

4. Rudolf Carnap, "The Elimination of Metaphysics through Logical

Analysis," in *Logical Positivism*, ed. A. J. Ayer (Glencoe, Ill.: Free Press, 1960), 60–81.

5. I use this term in the largely descriptive sense that it has, e.g., for K. O. Apel, who divides contemporary philosophy into two major currents: analytic and existential. See Apel's *Transformation der Philosophie* (Frankfurt: Suhrkamp Verlag, 1973).

6. Nietzsche, *Human All-Too-Human*.

7. Theodor W. Adorno, *Negative Dialectics*, trans. E. B. Ashton (New York: Continuum, 1973). Further references to this work, abbreviated ND, will be included in the text.

8. Translator's note: I have modified slightly the syntax of the English translation of this phrase from Adorno.

9. Translator's note: I have modified the English translation, which inexplicably translates the town names as "Applebachsville, Wind Gap, or Lords Valley."

10. See Theodor W. Adorno, *Aesthetic Theory*, trans. C. Lenhardt, ed. Gretel Adorno and Rolf Tiedemann (1970; rpt. London: Routledge & Kegan Paul, 1984).

11. Martin Heidegger, *Der Satz vom Grund* (Pfullingen: Neske, 1957).

12. Emmanuel Lévinas, *Totality and Infinity*, trans. Alphonso Lingis (Pittsburgh: Duquesne Univ. Press, 1969), 45. Further references to this work, abbreviated TI, will be included in the text.

13. Translator's note: Vattimo here uses *faccia* and *volto* to differentiate between the two meanings, and in the rest of the essay he uses the term *volto*, to refer to face. Although the English "visage" might be a more accurate translation of the Italian *volto*, and "face" a more accurate translation of *faccia*, I have followed the English translations of Lévinas, in which the French *visage* is rendered as "face."

14. The essay is significantly titled "Violence and Metaphysics: An Essay on the Thought of Emmanuel Lévinas." In Jacques Derrida, *Writing and Difference*, trans. Alan Bass (Chicago: Univ. of Chicago Press, 1978).

15. Ibid., 125ff.

16. Ibid., 147.

17. Ibid.

18. See Emmanuel Lévinas, *Humanisme de l'autre homme* (Montpellier: Fata Morgana, 1971), 40.

19. Emmanuel Lévinas, *Difficile liberté* (1963; rpt. Paris: Albin Michel, 1976), 230.

20. On this point see the comments by S. Petrosino and J. Rolland in *La vérité nomade: Introduction à E. Lévinas* (Paris: La Découverte, 1984), 73. This essay and Derrida's have been particularly present to me. Among the recent writings on Derrida in Italian, see also M. Ferraris's dense essay "L' esclusione della filosofia: Ebraismo e pragmatismo," *Aut Aut* no. 123 (1986), for important observations on the theme of secularization.

21. See the already cited pages from the preface of *Totality and Infinity*.

22. Emmanuel Lévinas, *Autrement qu'être ou au-delà de l'essence* (The Hague: Martinus Nijhoff, 1974), 3.

23. On the notion of *Verwindung* in Heidegger, see the last chapter of Gianni Vattimo, *La fine della modernità* (Milan: Garzanti, 1985). Also available in French translation as *La fin de la modernité* (Paris: Seuil, 1987).

Toward an Ontology of Decline

1. Martin Heidegger, *Nietzsche* (Pfullingen: Neske, 1961).

2. Martin Heidegger, *Identität und Differenz* (Pfulligen: Neske, 1957); *Identity and Difference*, trans. Joan Stambaugh (New York: Harper & Row, 1969).

3. Friedrich Nietzsche, *Human All-Too-Human*, Part 2, in *Opere*, ed. Giorgio Colli and Mazzino Montinari, Vol. 4 (Berlin: de Gruyter, 1964).

4. Hans Georg Gadamer, *Truth and Method* (New York: Seabury Press, 1975).

5. Martin Heidegger, *Zur Sache des Denkens* (Tübingen: Niemayer, 1969), 5–6.

6. See Martin Heidegger, *Being and Time* (New York: Harper & Row, 1962), par. 46.

7. Friedrich Nietzsche, *Human All-Too-Human*, Part 2, trans. Paul V. Cohn (New York: Russell & Russell, 1964), 239.

8. Heidegger, *Identity and Difference*, 38 (translation modified).

9. Ibid., 35–38.

10. Martin Heidegger, "The Thing," *Poetry, Language, Thought*, trans. Albert Hofstadter (New York: Harper & Row, 1971), 180–82.

11. Heidegger, *Identity and Difference*, 35.

12. Ibid., 37 (translation modified).

13. See ibid., 40.

14. See ibid., 32–33.

15. See Heidegger, "The Onto-Theological Constitution of Metaphysics," *Identity and Difference*.

16. Martin Heidegger, *Der Satz vom Grund* (Pfullingen: Neske, 1957), 186–87.

Venusian Charme

1. Jean Baudrillard, *De la séduction* (Paris: Galilée, 1979).

2. Pascal Bruckner and Alain Finkielkraut, *Le nouveau désordre amoureux* (Paris: Seuil, 1977).

3. A. Vergote, "Charmes divins et déguisements diaboliques," in *La séduction*, ed. M. Olender and J. Sojcher (Paris: Aubier, 1980).

4. R. Schilling, *La religion romaine de Vénus depuis les origines jusqu'au temps de Auguste* (Paris: De Boccard, 1954). See also the articles devoted to Venus collected in Schilling, *Rites, cultes, dieux de Rome* (Paris: Kliencksieck, 1979); as well as G. Dumézil, *Idées romaines* (Paris: Gallimard, 1969), which adds the term *venustas* to those listed by Schilling.

5. R. Radiguet, *Les joues en feu*, in *Oeuvres complètes* (1952; Paris: Slatkine Reprints, 1981); translated as *Cheeks on Fire* by Alan Stone (London: Calder, 1976).

6. G. Dumézil, *Déesses latines et mythes védiques* (Paris: Gallimard, 1956).

7. Diodorus Siculus, *Biblioteca historica*, trans. C. H. Oldfather, Loeb Classical Library (London: Heinemann, 1833–1967), 4.83.5.

8. Giambattista Marino, *L'adone* (Bari: Laterza, 1975–77), canto 20, line 92.

9. M. Olender, *Une magie de l'absence*, in *La séduction*.

10. R. Radiguet, *Le diable en corps* (Oxford: Blackwell, 1983); English translation, *The Devil in the Flesh*, by A. M. Sheridan Smith (Boston: M. Boyars, 1982).

11. See Mario Perniola, *La società dei simulacri* (Bologna: Cappelli, 1980), 180–83. An interpretation of the relationship between Don Juan and the statue close to my own is that of J. N. Vuarnet, "Le séducteur malgré lui," in *La séduction*, 72. The importance of the connection between Don Juan and death, which most interpreters overlook, is emphasized by J. Rousset, *Le mythe de Don Juan* (Paris: A. Colin, 1978).

12. D. De Rougement, *Love in the Western World*, trans. Montgomery Belgion (Philadelphia: Saifer, 1953).

13. Livy, *Ab urbe condita* 8.9.13. See also H. Fugier, *Recherches sur l'expression du sacré dans la langue latine* (Paris: Les Belles Lettres, 1963); and Fugier, *Temps et sacre dans le vocabulaire religieux des Romains* in "Archivio di filosofia," *Mito e fede* 1966.

14. G. Wissowa, *Religion und Kultus der Römer* (Munich: Beck, 1912), 289.

15. R. Radiguet, "Statue or Scarecrow" ("Statue ou épouvantail"), *Cheeks on Fire*, 70–71. English translation modified.

16. E. Benveniste, *Le vocabulaire des institutions indo-européenes* (Paris: Minuit, 1970).

17. P. Cipriani, *Fas e nefas* (Rome: Instituto di Glottologia dell'Università di Roma, 1978), 19.

18. This is a recurrent theme of Baudrillard's *De la séduction*. E.g., "Seduction is a destiny: in order for it to be fulfilled, all freedom must be there, but also wholly extended, like a somnambulist, toward its loss" (p. 147 of the French edition).

19. There is an echo of this definition in Bruchner and Finkielkraut's *Le nouveau désordre*: "Erotic short circuits emerge and upset acquired classifications from the inside."

20. Cicero, *De divinatione* 1.23.

21. Livy, 6.28.

22. See Angelo Brelich, *Tre variazioni sul tema delle origini* (Rome, 1955).

23. Valerius Maximus, 7.1.

24. Plutarch, *Sulla*, in *Plutarch's Lives*, the Dryden Plutarch revised by Clough (New York: Dutton, 1910), 2.175.

25. Appianus, *De bellis civilibis* 2.69.

26. Schilling, *La religion romaine de Vénus*, 315.

27. Edith Harrison, *Prolegomena to the Study of Greek Religion* (Cambridge, 1903), 95.

28. René Girard, *La violence et le sacré* (Paris: Grasset, 1972), 358; translated as *Violence and the Sacred* by Patrick Gregory (Baltimore: Johns Hopkins Univ. Press, 1977).

29. Jacques Derrida, "Plato's Pharmacy," in *Dissemination*, trans. Barbara Johnson (Chicago: Univ. of Chicago Press, 1981).

30. Schilling, *La religion romaine de vénus*, 133ff.

31. H. Jeanmaire, *Dionysos* (Paris: Payot, 1951), 23.

32. Girard, *La violence et le sacré*, 188–89.

33. Ovid, *Fasti* 3.345; and Plutarch, *Numa Pompilius*, in *Plutarch's Lives*.

34. Livy, 10.42.7.

35. See p. 2, chap. 3, of A. Bruhl, *Liber pater: Origine et expansion du culte dionysiaque à Rome et dans le monde romain* (Paris: De Broccard, 1953).

36. G. Baffo, "Venere e Adone," in *Poesie* (Milan: Mondadori, 1974).

37. Radiguet, "Statue or Scarecrow," 71 (translation modified).

38. Girard, *La violence et le sacré*, 408.

39. Plutarch, *Numa Pompilius*.

40. Such a conclusion is clearly at odds with the basic thesis of Schilling's book. For him, *venus* implies a total devotion to the deity that stands in opposition to *fides*, that is, the joint contract between human and supernatural. Romulus's religious attitude is thus an expression of *venus*, while Numa Pompilius would illustrate *fides*. *Venus* is emotional and magical-mystical, and has a dimension of interiority and supplication, whereas *fides* is rational and juridical and has a dimension of exteriority and

formalism. But doesn't the particular quality of Roman religion reside precisely in the *Verwindung* of these oppositions?

Decorum and Ceremony

1. E.g., in *Iliad* 12.104. See the most extensive study available: M. Pohlenz, *To prepon: Ein Beitrag zur Geschichte des greichischen Geistes* (1933), in *Kleine Schriften* (Hildesheim: Olms, 1965), 100–139.

2. The Indo-European root "prep-" means precisely "to fall under the glance, appearance, form." See J. Pokorny, *Indogermanisches Wörterbuch* (Bern and Munich, 1959), 1:845.

3. P. Chantraine, *Dictionnaire étimologique de la langue grecque* (Paris: Kliencksieck, 1968).

4. K. Kerényi, *Die antike Religion* (Munich and Vienna: Lagen-Müller, 1969).

5. Martin Heidegger, *An Introduction to Metaphysics*, trans. Ralph Manheim (New York: Doubleday, 1961).

6. Pindar, *The Pythian Odes*, trans. Lewis Richard Farnell (London: Macmillan & Co., 1930), 10.67.

7. No. 150 in Edgar Lobel and Denys Page, *Poetarum lesbiorum fragmenta* (London: Oxford Univ. Press, 1963).

8. A. Rostagni, "Un nuovo capitolo nella storia della retorica e della sofistica," *Studi italiani di filologia classica*, 2, nos. 1–2 (1922): 148–201.

9. Q. Cataudella, "Sopra alcuni concetti della poetica antica, I, *Apate*," *Rivista di filosofia classica* 59 (1931): 328–87.

10. Gorgias, in M. Untersteiner, ed., *Sofisti: Testimonianze e frammenti* (Florence: La Nuova Italian, 1967), 2:87.

11. In M. Untersteiner, *I sofisti* (Milan: Lampugnani Nigri, 1967), 1:251.

12. In Untersteiner, *Sofisti*, 2:143.

13. In ibid, 2:99.

14. *The Greater Hippias*, in *The Collected Dialogues of Plato*, ed. Edith Hamilton and Huntington Cairns, trans. Benjamin Jowett (1961; rpt. Princeton: Princeton Univ. Press, 1973), 291d, 292c-d.

15. Ibid., 294b.

16. Ibid., 294e.

17. Werner Jaeger, *Paideia: The Ideals of Greek Culture*, trans. Gilbert Highet (1947; rpt. New York: Oxford Univ. Press, 1963).

18. Aristotle, *Rhetoric* 1355a, 21. The English translations are those of W. Rhys Roberts (New York: Random House, 1954).

19. M. Pohlenz, *Antikes Führertum: Cicero De officiis und das Lebensideal des Panaitios* (Leipzig and Berlin: Teubner, 1934), 55ff.

20. Pohlenz, *To prepon*, 107 n. 2. In a passage of *De oratore*, Cicero translates *prepon* as *aptus*.

21. P. Monteil, *Beau et laid en latin: Etude de vocabulaire* (Paris, 1964), 72ff.

22. The translation used is that of H. M. Hubbell in the Loeb Classical Library (London: Heinemann, 1971).

23. K. H. Roloff, "Caerimonia," *Glotta: Zeitschrift für griechische und lateinische Sprache* 32 (1953): 101–38.

24. Kerényi, *Die antike Religion*.

25. G. Piccaluga, *Elementi spettacolari nei rituali festivi romani* (Rome: Editori dell'Ateneo, 1965), 64.

26. G. Dumézil, *La religion romaine archaïque* (Paris: Payot, 1974), 50.

27. Roloff, "Caerimonia," 111.

28. See Emmanuel Lévinas, *Totality and Infinity*, trans. Alphonso Lingis (Pittsburgh: Duquesne Univ. Press, 1969).

29. R. von Jhering, *Der Zweck im Recht* (Leipzig: Breitkopf & Härtel, 1904–5.).

30. Roloff, "Caerimonia," 121.

The Black Light

The pages that follow are part of a larger work published as *La posta in gioco: Husserl, Heidegger, il soggetto* (Gazanti, Milan, 1987). In particular, the first part of this work analyzes the contemporary "vicissitudes" of the Cartesian *cogito* with reference to Husserl, Heidegger, Foucault, Derrida, and Lévinas.

1. J. Derrida, "Cogito and the History of Madness" and "Violence and Metaphysics," in *Writing and Difference*, trans. Alan Bass (Chicago: Univ. of Chicago Press, 1978).

2. Derrida, "Cogito," 61.

3. Ibid.

4. Cf. H. Corbin, *The Image of the Temple* (1980). Corbin has also recently been referred to by F. Rella in *Metamorfosi* (Milan: Feltrinelli, 1984), 95ff.

5. Corbin, *The Image of the Temple*, 114. Corbin refers to the conversation of the sixth Imam, Ja'far al-Sadiq, with one of his disciples. His source is the volume by H. Bietenhard, *Die himmlische Welt im Urchristentum und Spätjudentum* (Tübingen, 1951), 231–54.

6. J. Derrida, "White Mythology," in *Margins of Philosophy*, trans. Alan Bass (Chicago: Univ. of Chicago Press, 1982), 213. Also, Derrida, "Le retrait de la métaphore," *Poésie* 6 (1979): 103–26.

7. In the sense analyzed by Derrida in "D'un ton apocalyptique adopté naguère en philosophie," in *Les fins de l'homme* (Paris: Galilée, 1981), the proceedings of a conference held at Cerisy-la-Salle in 1980. This article has been translated into English by John P. Leavey, Jr., as "Of an Apocalyptic Tone Recently Adopted in Philosophy," *Semeia* 23 (1982): 63–97.

8. Derrida, "White Mythology," 271.

9. One of the fundamental motifs of Lévinas's thought of alterity is precisely the interpretation of the idea of God and the infinite of Descartes: this motif circulates in all of Lévinas's work up until the recent *De dieu qui vient a l'idée* (Paris: Vrin, 1982).

10. R. Descartes, *Entretien avec Burman*, in *Oeuvres et lettres*, ed. A. Bridoux, 2d ed., Bibliothèque de la Pléiade (Paris, 1963), 1387–88. The passage is cited by Derrida in "White Mythology," 268.

11. E. Husserl, *Zur Phänomenologie des inneren Zeitbewisstseins* (1928), ed. R. Boehm (The Hague: Nijhoff, 1966).

12. J. Derrida, *Speech and Phenomena and Other Essays on Husserl's Theory of Signs*, trans. David B. Allison (Evanston, Ill.: Northwestern Univ. Press, 1973), 84, n. 9.

13. J. Derrida, *Writing and Difference*, 92. Blanchot objects: "The face—although I have to admit that the name constitutes a difficulty—is the presence which I cannot dominate with the gaze, which always transcends the

representation that I can make of it and every form, every image, vision and idea in which I can affirm it, arrest it, or simply let it be present," M. Blanchot, "Conaissance de l'inconnue," in L'entretien infini (Paris: Gallimard, 1969), 77.

14. See M. Blanchot, "L'oubli: La déraison," in L'entretien infini.

15. Of particular interest, as philosophic elaboration within this horizon, one should see M. Cacciari's recent considerations on the religious "icon" with reference to the writings of P. Florenskij and the painting by V. S. Malevič: see Icone della legge (Icons of Law) (Milan: Adelphi, 1985), 173ff. For example: "The mind concentrates the luminous principle, it takes back to its point where it reaches such an intensity as to no longer be able to 'free itself' from itself and, consequently, to no longer be able to produce the world of maya. That only is Light: the black hole. A formidable implosion sucks in all the sensible by way of the window of the icon amassing it in the invisible point where every direction, every sense and dimension simultaneously remain as possibles. The black hole is the term that is destined for that vortex of the invisible that the window of the icon produces" (208).

16. G. Bachelard, La terre et les rêveries du repos (Paris: Corti, 1948), 27–28. Compare, however, all of the previous chapter dedicated to the "reveries de l'intimité materielle." Derrida refers to this book by Bachelard in Writing and Difference.

17. G. Bachelard, La terre et les rêveries du repos, 20, 24.

18. Regarding the metaphor of the "black spot," see the essay by Derrida on Bataille in Writing and Difference.

19. See E. Lévinas, De dieu qui vient a l'idée, 51n.

20. See, e.g., the reference to the "come" as "voice" which is not reduced in any linguistic register, in the last part of "Living on/Border Lines," in Deconstruction and Criticism, by H. Bloom et al. (New York: Seaburg Press, 1979), 75–176.

21. J. Derrida, Writing and Difference.

22. Cf. in Lévinas, as terminal point of his critical excavation within phenomenology, the essay entitled "De la conscience à la veille," in De dieu qui vient a l'idée, 34–61.

The Atopy of the Modern

1. Simone Weil, Notebooks (Boston: Routledge & Kegan Paul, 1976), 230.

2. P. Florenskij, La colonna e il fondamento della verità, trans. P. Modesto (Milan: Rusconi, 1974), 194.

3. See V. Jankélévitch, L'ironie (Paris: Flammarion, 1964), 113ff., for a list of the numerous passages in which this term appears in relation to Socrates. The fundamental elaboration of this concept or figure of thought occurs in the Symposium. Quotes from L. Robin refer to the "Notice" that precedes his edition of Le Banquet (Paris: Les Belles Lettres, 1981). Robin translates "atopos" as "déroutant," misleadingly.

4. Friedrich Schlegel, Lucinde, in Kritische Ausgabe seiner Werke, ed. E. Behler et al. (Paderborn: Schöningh, 1958), 6:61.

5. Giacomo Leopardi, Zibaldone, ed. Francesco Flora (Milan: Mondadori, 1976). On the noetic force of the image, see H. Corbin, Corpo spirituale e terra celeste, trans. G. Bemporad (Milan: Adelphi, 1986).

6. Walter Benjamin, Gesammelte Schriften, ed. R. Tiedemann and H. Schweppenhäuser (Frankfurt: Suhrkamp, 1972–85). Friedrich Dürrenmatt, Dramaturgy of the Labyrinth. On the theme of the mutation of the fundamental metaphors, see H. Blümenburg, Paradigmen zu einer metaphorologie (Bonn: Bouvier, 1960).

7. L. Massignon, "Finisterre," in Nell'Islam: Giardini e moschee, 2 (1986). Schlegel, Kritische Ausgabe seiner Werke, 2:318–19, 2:313; vol. 18, sect. 4, frag. 471; vol. 18, sec. 2, frag. 592. The arabesque pose presents itself as an original form because it contains all possible forms. The very form of possibility is therefore, insofar as it is not yet realized, null. But this null is not nothing: it is rather an "infinite fullness."

8. Friedrich Schlegel, "Athanäeum," in Kritische Ausgabe seiner Werke, 1:80.

9. Friedrich Dürrenmatt, The Judge and His Hangman, trans. Cyrus Brooks (London: J. Cape, 1967).

10. K. K. Polheim, Die Arabeske, Ansichten und Ideen aus Friedrich Schlegels Poetik (Paderborn: Schöningh, 1966), 113.

11. F. W. J. Schelling, letter to Hegel, 4 Feb. 1795, in G. W. F. Hegel, Briefe von und an Hegel, ed. Johannes Hoffmeister, 4 vols. (Hamburg: Felix Meiner, 1969–81).

12. F. W. J. Schelling, Vom Ich als Prinzip der Philosophie oder über das Unbedingte im menschlichen Wissen (1795), in Schelling, Ausgewählte Schriften, ed. M. Frank (Frankfurt: Suhrkamp, 1985), 1:56.

13. Novalis, Werke, Tagebücher und Briefe, ed. H. J. Mäahl and R. Samuel

(Munich: Hanser, 1978). The fragments cited in the text are from vol. 2, p. 227, frag. 1; p. 226, frag. 304.

14. Ibid., 2:666, frag. 600. This fragment, entitled "Theory of Pleasure," could constitute the title of an erotics of knowledge, which is perhaps Novalis' secret project, the point of convergence in his immense fragmentary work, which can be compared only to Leopardi's *Zibaldone*. Frags. 124–25, vol. 2, p. 343, are also cited in the text.

15. Schelling, *Ausgewählte Schriften*, 5:666.

16. Ibid., 2:840.

17. I have in mind the painting of M. N. Rotelli, *Aperto*, Venice Biennale, 1986.

18. Novalis, *Werke*, vol. 2, *Nachlese*, pp. 20ff. Schlegel attempted at various times to define the category of the modern, but it was by then a characteristic of his age which affected Hegel as well.

19. It is not by chance that Baudelaire stands at the center of Benjamin's *Passagen-Werk*, another great attempt to grasp and represent this new category of thought. Baudelaire is not only one of the greatest—or the greatest—poets of the nineteenth century; his pages of art and literary criticism are also memorable, as are some of the theoretical affirmations scattered in the *Journaux intimes*.

20. V. Solov'ëv, *Il significato dell'amore e altri scritti*, ed. A. Dell'Asta (Milan: La Casa di Matriona, 1983); Emmanuel Lévinas, *Le temps et l'autre* (Paris: PUF, 1983).

21. Jean François Lyotard, *La condition postmoderne* (Paris: Minuit, 1979); Lyotard, *The Postmodern Condition*, trans. Geoff Bennington and Brian Massumi (Minneapolis: Univ. of Minnesota Press, 1984).

22. Jean François Lyotard, *Le postmoderne expliqué aux enfants* (Paris: Galilée, 1986), 25.

23. Ibid., 29–33. Habermas's positions are set forth in his 1980 Adorno Prize speech.

24. See Franco Rella, *La battaglia della verità* (Milan: Feltrinelli, 1986), chap. 4.

25. Søren Kierkegaard, *Papirer*, ed. P. A. Heiberg (Copenhagen: Kuhr and Torsting, 1909–48), vol. 2, A 603.

26. Giorgio Agamben, "La cosa stessa," in *Disegno*, ed. G. Dalmasso (Milan: Jaca Book, 1984), 3. The reference is to the theories of Gaiser and Krämer; see Hans Krämer, *Platone e i fondamenti della metafisica*, ed. G. Reale (Milan: Vita e Pensiero, 1982).

27. Or at least this may be the fear of the author. I don't think it is rare for a narrator, faced with certain aspects of his character, to operate what Schlegel called a "parabasis," that is, a sort of interruption of the narrative autonomy of the character itself in order to signal that that figure is not the container of his ideas or that, vice versa, it expresses things that do not belong to its author.

28. Solev'ëv, *Significato dell'amore*, 322.

29. Søren Kierkegaard, *The Concept of Irony with Constant Reference to Socrates*, trans. Lee Capel (London: Collins, 1966).

30. Kierkegaard, *Papirer*, vol. 4, A 68.

31. Ibid., vol. 5, A 68.

32. Ibid., vol. 2, A 113.

33. Paul Ricoeur, *Temps et récit* (Paris: Seuil, 1983–85), 391–92.

34. Schlegel, *Kritische Ausgabe*, vol. 18, sec. 4, frag. 1471.

35. Novalis, *Werke*, 2:684, frag. 906.

36. See Luigi Pareyson, *Lo stupore della ragione in Schelling*, in *Romanticismo, esistenzialismo, ontologia della libertà* (Milan: Mursia, 1979); Pareyson, "Filosofia e esperienza religiosa," *Annuario filosofico Mursia* (1985); Martin Heidegger, *Gesamtausgabe*, Pt. 1, *Veröffentliche Schriften*, vol. 9, and Pt. 2, *Vorlesungen 1923–1944*, vols. 29–30 (Frankfurt: Klostermann, 1976–83); Sigmund Freud, "The Uncanny," in *The Standard Edition of the Complete Psychological Works of Sigmund Freud*, trans. James Strachey, vol. 17 (London: Hogarth, 1953-74); Paul Valéry, *Oeuvres*, vol. 2 (Paris: Gallimard, 1977), 166. On Valéry, see Franco Rella, *Metamorfosi: Immagini del pensiero* (Milan: Feltrinelli, 1984).

37. Novalis, *Werke*, 2:619, frag. 633. "Chance and necessity (the external through me). . . . One is necessarily frightened when one casts a glance into the depths of the spirit. The sense of depth and the will have no limits. It is thus like the heavens. Exhausted, the power of imagination stands immobile."

38. Ibid., 2:771, frag. 138.

39. Peter Handke, Die Lehre der Sainte-Victoire (Frankfurt: Suhrkamp, 1980).

40. Novalis, Werke, 2:675, frag. 857.

41. See Schlegel's texts in Polheim, Die Arabeske, 29–31.

The Problem of Representation

1. Walter Benjamin, "Schicksal und Charakter," in Gesammelte Shriften, vol. 2, pt. 1 (Frankfurt: Suhrkamp, 1980), 175.

2. The theme (already of Schelling and Hölderlin) of the end of demonic necessity, of the Übermacht of destiny, also appears central in the Star of Redemption by Franz Rosenzweig. The figure of the "servant of God," simple subject of the judgment that destiny pronounces, is contrasted, for man and the world, with the "light of revelation." Gratia sua, the direction of the will is not demonically fixed once and for all, "but in every moment dies and in every moment it renews itself" (F. Rosenzweig, Der Stern der Erlösung, 2d ed. [Frankfurt: Suhrkamp Verlag, 1930], pt. 2, book 3, pp. 160-63; trans. as The Star of Redemption by William Hallo [New York: Holt, Rinehart, and Winston, 1971]). It would be of considerable interest in this respect, moreover, also to develop an analysis of the relationship—until now only touched upon—between Benjamin and Aby Warburg. "The struggle with the monster" ("der Kampf mit dem Monstrum"), the passing, extremely "periculosum" and never once and for all overcome, from the "monstrous complex to the ordering symbol" ("vom monstrosen Komplex aum ordnenden Symbol") (quoted in H. Gombrich, Aby Warburg [London: The Warburg Institute, 1970], 251–52, the ability to safeguard—save the past dominating its immediately demonic appearance—all the "polarities" of the genius of Warburg have profoundly to do with the Benjaminian concept of character. "Homo victor" one could say, with Warburg (see ibid., 322), he who "remains master of his forces: man enough to break the continuum of history" (Walter Benjamin, "Thesis on the Philosophy of History," in Angelus Novus, in Schristen [Frankfurt: Suhrkamp, 1955]).

3. F. W. J. Schelling, Philosophie der Kunst, in Sammtliche Werke (Stuttgart, 1856–61), vol. 1, pt. 5, p. 697.

4. G. Colli, La nascita della filosofia (Milan: Adelphi, 1975), 67.

5. The following Benjaminian citations are taken from Ursprung des deutschen Trauerspiels, in Gesammelte Schriften, vol. 1, pp. 203–ff., trans. as The Origin of German Tragic Drama by John Osborne (London: New Left Books, 1977).

6. On the distance between philosophy and "research" see the very interesting pages by M. Sgalambro in *La morte del sole* (Milan: Adelphi, 1982).

7. The absence of properly gnostic suggestions on Benjamin has been clearly indicated both by F. Desideri in *Il tempo e le forme* (Rome: Armando, 1980), and by G. Schiavoni in *Walter Benjamin: Sopravvivere alla cultura* (Palermo: Sellerio, 1980). In this respect the difference with the thought of Ernst Bloch is also evident.

8. It is probably on this problem (that of the symbolic-iconic sign) that Benjamin's constant interest is centered with regard to romantic aesthetics and criticism. Concerning this problem see T. Todorov in *Theories of the Symbol*, trans. Catherine Porter (Ithaca, N.Y.: Cornell Univ. Press, 1982). It is against the background of these romantic theories that it would prove to be necessary to face the relationship between Benjamin's concept of Eros and that of Klages (finally rescued from bad literary "spells," as managed by G. Moretti in *Anima e immagine: Sul "poetico" in Ludwig Klages* [Palermo: Sellerio, 1985]), which is expressed in the discussion of the soul with the image, an image which the soul does not derive from itself but from which it is struck, which it *undergoes*, as the image of distance, unable to be possessed.

9. This is one of the most profound affinities between Benjamin and Kafka: for Kafka, in fact, impatience constitutes *the* sin. Regarding this theme I refer to the second chapter of the first part of my book *Icone della Legge* (Milan: Adelphi, 1984), the chapter dedicated to the Kafkian "integration."

10. Rosenzweig, *Der Stern der Erlösung*, 210.

11. Cf. E. Przywara, "Zeit, Raum, Ewigkeit," in *Tempo e eternità* (Padua: CEDAM, 1959); and E. Grassi, "Apocalisse e storia," in *Apocalisse e insecuritas* (Milan and Rome, 1954).

12. For the general picturing of these motifs within the ambit of the Jewish mystical tradition and its radical discussion in Benjamin (and, in other respects, also in Kafka) the following are fundamental: G. Scholem, *Kabbalah* (New York: Quadrangle, 1974); and Scholem, *Zum Verständis der messianischen Idee im Judentum*, in *Judaica*, vol. 1 (Frankfurt: Suhrkamp Verlag, 1977).

Time and Alienation

1. Aristotle, *De Interpretatione*, trans. J. L. Ackrill (Oxford: Clarendon Press, 1974).

2. See the Acts of the Conference on "The Sacred," Centro Internazionale di Studi Umanistici, Rome, 1974, 72–73.

3. Mircea Eliade, *The Myth of the Eternal Return*, trans. Willard R. Trask (Princeton, N. J.: Princeton Univ. Press, 1954).

4. Acts of the Conference on "The Sacred," 73.

5. E. Severino, *Essenza del nichilismo* (Milan: Adelphi, 1982).

6. Ibid., 88–131.

The Earth and the Essence of Man

1. Translator's note: As opposed to that of Nietzsche.

2. E. Severino, "The Path of Day," in *Essenza del Nichilismo*, 145–95, 17, 21.

3. But also in the case of what is believed to appear, something is affirmed because it appears. One affirms, e.g., that also the unseen parts of a lamp exist. They certainly do not appear as the visible parts do (the specific error of naturalistic realism consists in the identification of these two ways of appearing): yet, in some way—and in any event according to a modality different from that of the visible parts—the nonvisible parts also appear precisely insofar as one speaks about them and is aware of them. It will be said that they appear as "ideal" determinations. So be it: but at the same time it is clear that, in this case as well, their existence is affirmed on the basis of the appearing of such a modality of existence.

4. See Severino, "Postscript" to "Returning to Parmenides," in *Essenza del Nichilismo*, 71.

5. G. W. F. Hegel, *Science of Logic*, trans. A. V. Miller (London: George Allen & Unwin, 1969), 85.

6. See, e.g., *ibid.*, 86.

INDEX

Abbagnano, N., 11
Abendland, 16, 63
Absence, 137
Adorno, T., 48–53, 60, 173
Aeschylus, 106
Aesthetics, 26, 30
Agamben, G., 3
Agathon, 105
Aletheia, 157. See also Truth
Alterity, 56–59, 133
Amor fati, 96
Anagkē, 175. See also Necessity
Ananke, 155
Anaximander, 22–23
Antirationalism, 14
Apel, K. O., 3, 65, 90, 148
Aporia, 4, 143
Appearance, 17, 50–51, 102; and Being,
    22, 106, 174–75, 179–81
Appolinaire, 102
Aquinas, T., 20–21
Arabesque, 138–39
Arche, 56, 60
Aristotle, 4, 20, 22, 24–25, 30, 52, 64,
    109–10, 122, 137, 145, 147, 150, 168,
    191–92; De doctrina christiana, 40;
    De interpretatione, 167; Metaphysics, 64,
    192; Physics, 22, 192; Poetics, 30;
    Rhetoric, 109–10
Atopy, 137, 142
Aufklärung, 8
Augustine, 7, 8, 14, 19, 30, 98
Austin, J. L., 31
Author, 26–43

Bachelard, G., 132
Baffo, G., 102
Balzac, 140, 146
Barthes, R., 29
Bataille, G., 128
Baudelaire, C., 32, 141
Baudrillard, J., 13
Beauty, 8, 105–7
Beckett, S., 99
Becoming, 4, 21, 172
Being, 3–4, 16, 33, 45, 48, 52–55, 59–61,

63–75, 180, 193–94; and Appearance,
    22, 106, 174–75, 179–80, 190–92; and
    non-Being/Nothing, 20–25, 167–75,
    177–80; and Truth, 67, 190–92, 181,
    183–84, 191–97
Benjamin, W., 50, 72, 138, 155–65
Benveniste, E., 98
Bergson, H., 7, 10
Blanchot, M., 132
Boehme, J., 148
Bonaparte, M., 39
Bonhoeffer, D., 172
Bontadini, G., 20–21
Booth, W., 28
Borges, J. L., 40, 130–31
Brelich, A., 99
Bruno, G., 147
Bultmann, R., 172

Cacciari, M., 4
Caesar, 113–14
Canaletto, 78, 80–81
Carchia, G., 3
Care, 18
Carnap, R., 47
Carrà, 5
Castelvetro, 150
Ceremony, 105–116
Cezanne, 140
Charme, 93–104
Chatman, S., 29
Chrysippus, 110
Cicero, 110–13
Cogito, 123, 130, 133, 147, 189
Concealment, 105
Conceptual schemes, 77–82
Conrad, J., 89
Consciousness, 7, 11, 86–88, 119–20,
    179, 183
Constructivism, 80
Corbin, H., 126
Corti, M., 29
Croce, B., 1, 7, 8–13, 20, 25

Dal Lago, A., 3
Dalmasso, G., 3

Death, 16, 22
De Chirico, 5
Decline, 18
Deconstruction, 5, 19, 22, 27, 34,
    41–42
Decorum, 18, 105–116
Deleuze, G., 3, 13, 17
Derrida, J., 3, 5, 14–17, 34–36, 38–39,
    41–43, 101, 120, 123–35; and Lévinas,
    55–60; *Le facteur de la vérité*, 38–39;
    Of Grammatology, 36; "Signature Event
    Context," 34–36; *Speech and Phenom-
    ena*, 129; *Writing and Difference*, 127
Descartes, R., 5–7, 20, 55–58, 78–81,
    123–35, 147–49
Desire, 55, 86
Destiny (*Geschick*), 3, 17, 22–23, 61, 70,
    155–56, 171, 179
Destruction, 47, 168
Dewey, J., 10
Dialectic, 142
Différance, 5
Difference, 113; ontological, 3, 14–15
Dilthey, W., 9–11, 90
Diogenes Laertius, 110
Dionysus, 101, 158
Dissemination, 15
Distance, 13, 18, 117–22, 124
Djik, T. V., 29
Doxa, 156–7
Doyle, C., 143
Dualism, 18–20
Dumézil, G., 95
Dürftiger Zeit, 68
Dürrenmatt, F., 138–39
Duty, 110

Earth, 23–24, 177–97
Eco, U., 18–20; *La struttura assente*, 30;
    Opera aperta, 31–32; *The Role of the
    Reader*, 31–32; *A Theory of Semiotics*, 31
Ego, 80–82, 89–90, 139, 141–2, 147
Eidos, 129, 159
Eliade, M., 172
Ellul, 171, 174
Empiricism, 5
Energeia, 52, 64
Epoché, 130, 133
Eros, 93–94, 138–39, 161. See also Love
Error, 3, 147, 183–84

Eschatology, 56–61
Ethics, 26, 53
Evil, 11, 93, 152–53
Excess, 123, 125–26
Existentialism, 19–20, 25

Fabbri, P., 30
Fabula, 39, 147–53
Face, 130
Fascism, 9, 20
Fichte, J. G., 139
Fillmore, C., 29
Finitude, 7, 10, 14
Florenskij, 137
Forster, E. M., 29
Fortune, 99
Foucault, M., 13, 29, 123, 126
Freedom, 22–23, 156
Freud, S., 45, 123–25, 145
Friction (*Attrito*), 13
Fromm, E., 174

Gadamer, H. G., 3, 7, 14–15, 29–30, 64, 66
Galileo, 78, 80–81
Galsworthy, J., 89
Galuppi, P., 8
Gargani, A., 3, 14, 20
Genette, G., 29
Gentile, G., 9
Ge-Stell, 47, 61, 71–74
Gilson, E., 20
Gioberti, V., 8
Girard, R., 101
God, 55–60, 114–15, 127–29, 134, 164,
    172–73
Goethe, J. W. V., 6
Goodman, N., 78, 80
Gorgias, 106–109
Gramsci, A., 1, 11, 14
Ground (*Grund*), 46–47, 52, 56, 59–60,
    65, 71, 172

Habermas, J., 65, 141, 171, 174
Hamann, J. G., 148
Hand, 22
Handke, P., 146
Hegel, G. W. F., 9–11, 51–52, 98, 113,
    128, 142, 183, 191–92; unhappy
    consciousness, 11
Heidegger, M., 3–4, 7, 12, 14, 16, 20–21,

47–48, 52–61, 63–75, 125, 130, 144,
151, 171, 173; *Befindlichkeit*, 66, 69;
Being and Time, 4, 47–48, 65, 70, 73, 75;
Dasein, 66–69, 71; *Er-eignis*, 71–74;
Geviert, 72; *Geworfenheit*, 67;
Grundbegriffe der Metaphysik, 145;
Indentität und Differenz, 63, 71–74;
Kehre, 70; *Nietzsche*, 63; "The Thing,"
72; *Wegmarken*, 145; *Zur Sache des
Denkens*, 73
Heraclitus, 175
Herbart, J., 10
Hermeneutic Circle, 19, 66
Hermeneutics, 14–20, 25–27, 30, 64–65;
and historicism, 4–6; and meta-
physics, 3–4
Hippias, 108
Hirsch, E. D., 29
Historicism, 5, 9, 10, 14, 20, 25
History, 6–7, 25, 59, 163, 172; of Being, 63
Hubris, 95–97
Humanism, 8
Husserl, E., 12, 30, 57, 129–30, 133–34

Idealism, German, 6, 9
Identity, 33, 72, 102, 124, 127–28, 170,
190
Imitation, 149
Infinite, 8, 54–55
Ingarden, R., 29
Intentionality, 12–13, 19
Interpretation, Theory of, 27–43
Iser, W., 29, 30, 37; *The Act of Reading*, 29,
31; *The Implied Reader*, 29
Isocrates, 108–110

James, H., 29, 37
Jauss, H. R., 29, 31, 32
Joyce, J., 29, 31; *Finnegans Wake*, 29;
Ulysses, 38
Jung, C., 145

Kafka, F., 50, 142, 144
Kant, I., 6, 9, 12, 30, 50–52, 65, 66, 74, 90,
129, 159; *Critique of Judgment*, 70
Kepler, J., 147
Kerényi, K., 113
Kierkegaard, S., 14, 142–44
Klossowski, P., 45
Knowledge, 6, 54, 65, 79, 153

Kojéve, A., 10
Kokoschka, O., 140
Krisis, 158–9
Kristeva, J., 29

Labriola, A., 11
Lacan, J., 120–22, 125, 151
Lack, 121–22
Language, 2, 13, 15, 33, 64, 107, 125,
129–30, 134, 169; and Being, 55;
and metalanguage, 41; of meta-
physics, 3, 46
Lawrence, D. H., 132
Leopardi, 138
Lévinas, E., 3, 14–15, 48, 53–61, 123, 126,
129–30; *Otherwise than Being*, 60;
Totality and Infinity, 54–58
Lévi-Strauss, C., 32
Light, 123, 129
Limit, 119, 142
Logic, 87, 150–51, 156, 191–92
Logos, 53, 57, 60, 110, 151, 155–56
Lotman, J., 29
Love, 93–94, 137–41
Lubbock, P., 29
Lukàcs, G., 15
Lyotard, J. F., 3, 141

Marc, F., 140
Marcel, G., 14
Marcuse, H., 171, 174
Margin, 18
Marino, G., 96
Maritain, J., 20
Marx, K., 51, 168; *Capital*, 11
Marxism, 1, 9–17, 45, 173–74; and
Ideology, 15
Mazzini, G., 8
Meaning, 13, 15, 89, 130, 173; of Being,
22, 24, 65, 173
Memory, 15, 18–19, 26
Merleau-Ponty, M., 12, 31
Metaphor, 117–122, 124–35, 151–53
Metaphysics, 3, 45–62, 63, 121, 126–29,
133, 177, 185, 189; of Presence, 129;
overcoming of, 3, 13, 23, 46–48, 53;
recoding, 1–26
Metonymy, 121
Miller, J. H., 40
Mind, 85

Mnemotechnics, 18
Model Reader, 19, 27–45
Modernity, 5, 23, 137–46, 153, 171
Morris, C., 30
Mukarovsky, J., 29, 30
Musil, R., 167
Myth, 6, 151

Nazism, 20, 48
Necessity, 22–26, 158, 175
Neo-Kantianism, 10
Neo-Mannerism, 5
Neo-Platonism, 8, 23, 121, 147
Neopositivism, 10, 12, 20
Neo-Scholasticism, 20, 25
Nietzsche, F., 3, 20, 45–47, 52, 60, 66, 70, 74, 123, 125, 128, 144; *Beyond Good and Evil*, 45; *Gay Science*, 17; Death of God, 46, 69, 73; *Human, All Too Human*, 17; *The Wanderer and His Shadow*, 45, 69; *Will to Power*, 47, 66, 68; Zarathustra, 76
Nihilism, 22–26, 66, 169–75, 178, 187
Nohl, H., 11
Nothing, 20–25, 122, 168–75, 177–79, 186–92
Noumenon, 20
Novalis, 139–40, 142, 145–46

Ockham, 18
Ontological Difference, 14
Ontology, 14–15, 53–57, 63–76, 172; fundamental, 65
Ontotheology, 63
Other, 54–55, 113, 183

Paci, E., 11–15, 20
Paideia, 108
Panaetius, 110
Paradigms, 77
Pareyson, L., 8, 11–15, 20, 31–33
Parmenides, 4, 20, 23, 24, 156–57, 174, 178, 190, 192
Passivity, 134–35
Patrizzi, 150
Peirce, C. S., 18, 30, 40–41
Perniola, M., 3, 17, 20
Personalism, 14
Petöfi, J., 29
Pharmakon, 100–103. *See also* Venom

Phenomenology, 11–15, 57, 133–34, 175, 183
Phenomenon, 20, 65
Philosophy, 1, 14, 26, 77, 124, 134, 145, 146, 161, 174, 181–82; Italian, 1–26; and Truth, 24; Speculative, 14
Phusis, 188–89
Piaget, J., 31
Pietas, 8
Pindar, 106
Plato, 3, 4, 14, 23, 20–25, 75, 101, 108, 121, 126, 128–29, 145, 148, 150, 155, 159, 183, 186, 190, 192; *Cratylus*, 151, 157–59; *Ion*, 108; *Laws*, 143; *Phaedo*, 143, 146, 177; *Phaedrus*, 153, 181, 184; *Republic*, 106, 143, 146–50, 153, 188; *Sophist*, 24, 178; *Symposium*, 137, 153; *Timaeus*, 147
Plotinus, 7
Poe, E. A., 38–39
Poetry, 7; and truth, 7, 148–53
Poiein, 178
Popper, K., 41, 47
Porphyry, 7
Positivism, 9, 45
Pouillon, 29
Pragmatism, 10, 37–38
Presence, 3, 129–30
Prepein, 105–112
Proust, M., 49
Prudentia, 114
Pugliatti, 30

Radiguet, R., 102
Reader, 19, 27–43
Reason, 16–17, 137, 173
Rella, F., 8, 20
Representation, 124, 155–65
Richards, I. A., 31
Ricoeur, P., 3, 14, 120–22, 171, 173–74
Riffaterre, M., 29
Rigobello, A., 3
Rilke, R. M., 132, 161
Robin, L., 137
Roloff, K. H., 112
Romanticism, 14
Rorty, R., 37–38
Rosenzweig, F., 164
Rosmini, A., 8
Rousseau, J., 6

Rovatti, P. A., 3, 20

Sacred, 113–14, 173
Same, 55
Savinio, 5
Scaligeri, 150
Schelling, F. V., 14, 139, 145, 155
Schilling, R., 94
Schlegel, F. V., 137–38, 140, 142, 144–46
Schleiermacher, 90
Schmidt, 29
Scholasticism, 20, 164
Science, 7
Scientism, 47
Scienza nuova, 6, 7
Searle, J., 31, 34, 36
Secularization, 45–61
Semantics, 29, 31, 36
Semiotics, 18–19, 27, 36, 42
Severino, E., 4, 20–24
Sexuality, 83, 93
Sign, 5, 13, 15
Simmel, G., 10, 72
Simplicius, 22
Simulacra, 13, 17
Sini, C., 3
Skinner, B. F., 173
Solov'ëv, 141
Sophists, 30
Space, 137
Speech-Act Theory, 31
Spiritualism, 8, 20, 25
Structuralism, 18–19, 30
Subjectivity, 7, 10, 12, 81–91, 125–26, 129–35, 146
Sublime, 30
Synthesis, 137, 139, 192

Technology, 3, 17, 25, 72–73, 173, 177–78, 180
Tekhnē, 178
Thing, 139, 156, 160
Thought, 26, 81, 84, 117; Weak, 6, 16
Thresholds, 18

Thucydides, 106
Time, 4, 18, 21, 115, 167–75, 170–72
Todorov, T., 29
Totality, 22, 54–57, 183, 185
Transavanguardia, 5
Transcendence, 52, 93
Transcendental, 65–66, 69
Transcendental Ego, 81–82, 89–90
Trendelenburg, F. A., 191
Truth, 7, 24, 37, 45, 58, 67, 79, 127, 159–60, 175; poetic, 7, 148–53; of Being, 181, 183–84, 191–97

Unconscious, 19–26, 123

Valery, P., 33, 147
Value, 66
Van Gogh, V., 78, 80, 81, 139
Vattimo, G., 3, 8, 16, 18, 20, 142
Veneratio, 94–97
Venerium, 99–100
Venia, 97–99
Venom, 100–103. *See also Pharmakon*
Vergil, 98
Vico, G., 5–8, 16, 19, 25, 148–53
Violence, 45–62, 125
Voltaire, 17
Von Baader, B. F. X., 148

Wahl, J., 10
Weakness (*Debolezza*), 13
Weber, M., 10
Weil, S., 137
Weinrich, 29
Whitehead, A. N., 12
Will, 156
Wittgenstein, 80
World, versions of, 77, 82

Xenophon, 107

Zeno, 110
Zola, E., 146

RECODING METAPHYSICS